Thomas Lamb Phipson

The Utilization of Minute Life

Being practical studies on insects, Crustacea, Mollusca, worms, polypes, Infusoria, and sponges

Thomas Lamb Phipson

The Utilization of Minute Life
Being practical studies on insects, Crustacea, Mollusca, worms, polypes, Infusoria, and sponges

ISBN/EAN: 9783337095314

Printed in Europe, USA, Canada, Australia, Japan

Cover: Foto ©berggeist007 / pixelio.de

More available books at **www.hansebooks.com**

THE UTILIZATION OF MINUTE LIFE.

THE

UTILIZATION OF MINUTE LIFE;

BEING

PRACTICAL STUDIES

ON

INSECTS, CRUSTACEA, MOLLUSCA, WORMS, POLYPES,
INFUSORIA, AND SPONGES.

BY

Dr. T. L. PHIPSON, F.C.S. London,

Late of the University of Bruxelles; Member of the Chemical Society of Paris; Laureate of the Dutch Society of Sciences; Corr. Memb. of the Belgian Entomological Society, the Pharmaceutical Society of Antwerp, the Society of Medical and Natural Sciences of Bruxelles, the Society of Sciences of Strasburg, etc., one of the Editors of " Le Cosmos," etc., etc.

Circular Coral Island, recently formed in the Pacific Ocean, principally composed of the species *Madrepora muricata*, and shutting in a portion of the ocean as a lake.

LONDON:
GROOMBRIDGE AND SONS.
MDCCCLXIV.

HARRILD, PRINTER, LONDON.

TO

WILLIAM SCHOLEFIELD, ESQ., M.P.,

ETC., ETC., ETC.

PERMIT me, my dear Sir, to dedicate this little volume to you, as a new proof of the high esteem in which I hold the *practical* efforts that have characterized your labours in Parliament, and of the personal friendship I bear to yourself.

Yours very sincerely,

THE AUTHOR.

PREFACE.

A VERY few words will suffice to make known my object in writing the present work.

Zoology and Botany have been looked upon as constituting less *practical* branches of Science than Chemistry or Astronomy, for instance. The zoological works placed in the hands of students are necessarily so full of anatomical details, details of classification, and observations upon the habits and instincts of animals, that very little space has (or could have) been afforded to notice the wonderful manner in which certain animals contribute *directly* to the welfare of mankind, and the methods by which they may be *cultivated*.

This remark is especially applicable to the lower classes of animals, to the *Invertebrata*, and to these I have devoted the following pages. Their investigation in a *practical* point of view has led, and will still lead, to very profitable and interesting results. It has been rendered more interesting of late years by numerous experiments, having for object the *culture and artificial propagation* of several of the more valuable species.

It is not sufficient to know that such an insect or such a polype is utilized for certain purposes in the Arts and Manufactures, we must acquire at the same time a correct idea of the animal itself, and the position it occupies in the animal kingdom; moreover, we must ascertain by experiment whether any species already valuable in its natural state cannot be rendered more so—cannot be submitted to *culture*, and *propagated* more extensively by artificial means, and thereby increase the benefits we derive from it.

To exhibit the actual state of this interesting question is the task I have imposed upon myself in the present work, which embraces the practical history of a great number of animals, and from which I find it impossible to exclude even the microscopic Infusoria.

When opportunity has been afforded I have mentioned a few peculiarities observable in several species, for it has been my endeavour to render the following pages interesting to the general student, as well as to the practical zoologist.

London, *January*, 1864.

CONTENTS.

CHAPTER I.

INTRODUCTION.

Domestication—Characteristics of a Species—Creation of Races and Varieties—Lost Types of the Animal Kingdom—Modified Species—Domestic Animals of Inferior Orders—Pisciculture—Creation of New Races of Fish—Cultivation of the Lower Animals . . . 1—8

CHAPTER II.

SILK-PRODUCING INSECTS.

Chemical Nature of Silk—The Spider's Web—Bombic Acid—Detection of Wool in Silk—Great Variety of Insects producing Silk—The Common Silkworm, *Bombyx mori*—The Golden Tree—The Province of Seres and the Morea—Prolongation of Life in Plants and Animals—Artificial Incubation and Rearing of *Bombyx mori*—Enormous Appetites—Insects living without Food—Rate at which the Silkworm spins—Modes of Destroying the Chrysalis—Calculation basis of Silk-breeding—The two Mulberry Trees—Diseases of Silkworms and their Remedies—Improvement of *Bombyx mori*—Tussah Silkworms—*Bombyx pernyi* and *B. Mylitta*—*Bombyx Cynthia*—Extraordinary Qualities of Silk—Other New Species of Silkworm—Spreading of these New Races—The Madagascar Silkworm—Production of Coloured Silk by the Insects themselves—Experiments—*Bombyx madróna*—Silk of the Clothes-Moth, *Tinea*—The Paraguay Spider—Ichneumon of the West Indies—Silk Imported into Liverpool

9—35

CHAPTER III.

COLOUR-PRODUCING INSECTS.

The Kermes—Latreille and his genus *Coccus*—*Coccus ilicis*—Crimson of the Romans—Brussels and Flemish Tapestries—*Coccus polonicus*—Coccus of the Poterium—*Coccus urva-ursi*—The Cochineal, *Coccus cacti*—Plants on which the Cochineal lives—Nopaleries—Grana sylvestra and Grana fina—Rearing of Cochineal—The Cochineal at Teneriffe—The Bluebottle Fly and the Aphides—Generation extraordinary—Two New Cochineals in Australia—*Cocus fabæ* (or Aphis fabæ) in France—Its Peculiar Colouring Matter—Lac—Carminium, its Discovery and Properties—The Colouring Matter of the Cochineal discovered in the Vegetable World—Carmine—Influence of Light in the Manufacture of Colours—*Rouge* for the face—Ink—The *Cynips*—Caprification—Diœcious Plants—Ripening of Figs in the East—Gall-nuts—*Cynips gallæ tinctoriæ*—Theory of the Formation of Vegetable Tumours—Analysis of Gall-nuts—Their Products and Uses—*Cynips quercus folii*—On the Formation of Grease by Animals—Other Insects Producing Dyes—*Aphis pini*—Money-spiders—The Magenta Dye and Cochineal . . . . 37—64

CHAPTER IV.

INSECTS PRODUCING WAX, RESIN, HONEY AND MANNA.

Chinese Coccus which produces a kind of Spermaceti—Value of its Produce—White Lac—Insects producing Resin—Wax Insect of Sumatra—Details concerning the wax Coccus—Bees—*Apis mellifica*—Its native country—Virgil—Modern Authors who have Written on Bees—*Apis ligustica*—*A. amalthea* and its curious Nests—Bamburos—*Apis unicolor*—Green Honey of Bourbon—Rock-honey of North America—*Apis fasciata*—*A. indica*—*A. Adansonii*—A Swarm of Bees—The Queen, Males and Workers—Mathematics of the Bee-cell—Silk produced by Bees—Production of Wax—How Honey is procured—Plants favourable to Bees—Duration of Life in Bees—Enemies and Maladies—Chloroforming Bees—Mr. Nutt's Hives—Profit derived from Bee-culture—New modes of Preserving Bees during Winter—Periodical Transportation of Hives—How to discover Bees' Nests—New Species of Bee at Sydney—Bees as Instruments of War—Honey, its Nature and Composition—Artificial Honey from Wood, Starch, etc.—Manna and the *Coccus maniparus*—Wax, its Nature, Composition, and Uses . . . 65—90

CHAPTER V.

INSECTS EMPLOYED IN MEDICINE, OR AS FOOD, AND OTHER INSECTS USEFUL TO MAN.

Spanish Flies—*Cantharides*—Their Medical Properties—Cantharidine—Cantharides in Poitou—Different Species of Cantharides—Discovery of Cantharidine in *Meloë*—The *Meloë*, or Oil Beetle—Metamorphoses of Moloë and Sitaris—*Cetonio aurata*—*Coccinella*—Trehala—*Buprestis*—Ants—Formic and Malic Acids in Ants—Production of Milk from the Eggs of Ants—Ants which collect Precious Stones—*Termes* as an Article of Food—Locusts and *Cicadæ*—*Acrydium migratorium*—The Ethiopian Acrydophagi—*Cicada septemdecim*—Bugs and Fleas—Southey—Phtirophagi—*Aranea edulis*—Centipedes—The Mexican Boat Flies—Beetle used for Soap—*Calandra granaria*—Presence of Tannic and Gallic Acids in this Beetle—Fire Flies—Truffle Flies—The Common House Fly, etc.—Remarkable Action of

Light upon Animal Life—Growth of Insects under differently Coloured Light . . . . 91—110

CHAPTER VI.

CRUSTACEA.

Artificial Propagation practicable with Crustacea as with Fish—The Common Lobster—Laws of Regeneration—The Craw Fish—Curious Discoveries relating to the Young of these Animals—*Phyllosoma*— *Zoëa*—Metamorphoses among Crustacea—*Praniza* and *Ancea*—Larvæ of Lobsters—Colouring Matter of Lobsters, Crawfish, etc.—Composition of a Lobster Shell—Shrimps —*Crangon vulgaris* —*C. boreas* —*Sabinea septem-carinata* and other Shrimps—Prawns — *Palemon carinus* and *P. jamaicensis* — Other Prawns—*Bopyrus crangorum* — The *Isopoda* — The Family of Crabs — *Cancer pagurus*—*C. mœnas*—*Pinnotheres*—*Pagurus*—Diogenes—Land Crabs—*Thelphusa fluviatilis*—Crabs of the genus *Gecarcinus*—Their Wonderful Emigrations—Bernardin de St. Pierre—*Birgus latro*—Robber Crab—Quantity of Fat it Produces—Concluding Remarks on this Family . . . . 111—134

CHAPTER VII.

MOLLUSCA.

CEPHALOPODA:—India and China Ink—Fossil Ink-bags—*Octopus vulgaris*—The Colour *Sepia*—*Sepia officinalis*, or

Cuttlefish—Cuttle-bone—*Loligo vulgaris*—Edible Cuttlefish—Chemical nature of their Colour—*Nautilus*—*Argonauta*—*Carinaria*.

GASTEROPODA :—The Tyrian Purple—Curious Properties of the Colouring Matter of Sea-snails—*Murex brandaris*—*Purpura lapillus*—*Helix fragilis*—*Yandina fragilis*—*Purpura patella*—*Murex truncatus*—Experiments with American Sea-snails—Colour furnished by Whelks—*Buccinum*—Influence of Light upon the Production of their Colour—Process used by the Ancients to dye Purple—Uric Acid in Gasteropoda—Murexide—Snails that are Reared for Food, etc.—*Helix pomatia*—Snail-gardens—*H. aspersa*—*H. horticola*—*Arion rufus*—Analysis of Snails—Limacine—Helicine—Uric Acid in *H. pomatia*—*Turbo littoreus*, or Periwinkle—Haliotis—Snails used as Money—*Cypræa moneta*—Other Species of Cypræa—"Love-shells"—*Conus*—*Oliva*—*Ovula*—*Strombus gigas*—Cassis—Turbinella—*Murex*—*Buccinum*—Curious Experiments with Snails—Slugs—*Limax maximus*—*L. agrestis*.

BIVALVES :—*Mytilus edulis*, or Common Mussel—Its Culture, etc.—Hurtful at certain seasons—*M. choros*—*M. Magellanicus*—*M. arca*—*M. lithophagus*—*Ostrea edulis*, or Common Oyster—Details concerning its Artificial Breeding and Propagation—Acclimatisation of Mollusca—Fishing on the Plessix bed—*Spondylus*—*Cardium edule*, or Cockle—*Solen*—*Pecten maximus*—*Tellina*—*Tridacna gigas*—*Chama*—Cameos—Stone Cameos and Shell Cameos—Chinese Cameos—Pearl Oysters—

Avicula margaritifera—A. frimbriata—A. sterna—Pearl Fishery—Details, etc.—Pearls of *Mytilus edulis—Anodontes—Unio pictorum—Unio margaritiferus*—Culture of the Fresh-water Pearl-Mussel—Value of its Pearls—Artificial modes of causing it to produce Pearls—*Pinna*—Their Silky Byssus and its uses—Their Pearls—Other uses of Shells—*Tunicata* and *Bryozoa* . . 135—198

CHAPTER VIII.

WORMS.

Curious Observations upon Worms—Reproductive Power of the *Naïs—Sabularia—Terebella—Lumbricus—Planaria*—Helminthes, or *Entozoa*—The Common Earth-worm, *Lumbricus terrestris*—The Leech, *Hirudo medicinalis*—The Horse-leech, *H. sanguisuga*—" Hirudiculture," or Leech-breeding—Its Cruelties—Extent to which it is carried in France—Barometers of Leeches and Frogs—Worms for the Aquarium . . . . 199—210

CHAPTER IX.

POLYPES.

General Remarks on Polypes—Their Organization and Polypidom—Naturalists who have written upon Polypes—*Hydra fusca* and *H. viridis*—Reproduction of Polypes—Polypes for the Aquarium—*Corallium nobilis* and general Observations on Coral—Its Polypidom—Practical Details concerning Coral — " Coralliculture "— Coral

Fishery—Uses of Coral—*Isis hippuris*, or Articulated Coral—*Tubipora musica*—The genus *Madrepora*—Reef and Coral Islands—Formation of Reefs—*Madrepora muricata*—Its Chemical Composition—How it derives its Lime—Its uses . . . . . 211—234

CHAPTER X.

INFUSORIA AND OTHER ANIMALCULÆ.

Microscopic Animals useful to Man—Universal Distribution of Infusoria—Dry Fogs—Authors who have studied Infusoria — Philosophical considerations concerning them — The *Monads, Rotifera, Vibrio* — *Rhizopoda*—*Monas crepusculum*, the most minute of living beings—Deposit in which the Transatlantic Cable lies—Transition of Colour in Lakes—Fossil Infusoria—Mountain Meal—Its Chemical Composition—Enormous quantities of it Consumed as Food—Geographical distribution of Infusorial Deposits—The Town of Richmond in Virginia—Berlin—The Polishing Schist of Bilin—1,750,000,000 beings to the square inch—The Swedish Lake Iron-ore—Tripoli, its uses and composition—Geographical and Geological Distribution of Infusoria, Foraminifera, and Diatomaceæ—Soluble Glass obtained from Infusorial Deposits—Its Uses—Other applications of Infusorial Earth—Chalk, its Uses and Origin—The Nummulite Limestone — Paris mostly built of Animalculæ—Other details—Time . . . 235—264

CHAPTER XI.

SPONGES.

Remarks on Classification—Structure of a Sponge—Naturalists who have contributed to the History of Sponges—Chemical Nature of Sponge—Interesting results—*Spongia officinalis* and *S. usta* — The Syrian toilet Sponge—Its high price—Other Sponges—Objects for the Aquarium—*Spongilla fluviatilis* and *S. lacustris*, or the Fresh-water Sponges—Sponges common on the English Coasts—Their use in Medicine—Sources of Iodine and Bromine—Flints and Agates as owing their formation to Sponges—Petrified Sponges—Practical details on the Toilet Sponge—Sponge Fishery and Markets . . . . 265—282

Chapter I.

Introduction.

Domestication—Characteristics of a Species—Creation of Races and Varieties—Lost Types of the Animal Kingdom—Modified Species—Domestic Animals of Inferior Orders—Pisciculture—Creation of New Races of Fish—Cultivation of the Lower Animals.

THE
UTILIZATION OF MINUTE LIFE.

INTRODUCTION.

THE lower classes of animals which are treated of in the following pages are mostly as remarkable for their great utility to man, as by the peculiarity of their organizations or their habits. Many of them have acquired as great an importance in the economic applications of the human race as the higher organized beings that have contributed to the welfare and comfort of man from the earliest historic periods, and which have generally been termed "domestic animals."

Such a term might, at the present day, be applied to most of those lower forms of animal life which will occupy our attention here.

By *domestication* is understood the art of training animals to administer to the wants of man. It is by flattering their natural tastes, by placing them artificially in circumstances similar in many respects to those of the savage state, preserving as much as possible their natural instincts, that the subjugation and domestication of the most useful species has been accomplished. It is still a discussed point among philosophers whether man has the power of modifying the nature of a species to such an extent that it loses its natural or essential characteristics.

However much the enthusiastic naturalist may admire the poetic doctrines of Lamarck, Etienne Geoffroy St. Hilaire, and Darwin, he must not completely throw aside Cuvier's more severe doctrine of the Fixity of Species. Both are true to a certain extent, but both have been exaggerated.

Domestic animals, like certain useful plants, have certainly undergone marked changes. No one doubts our power of creating new races or varieties in the animal world, with almost as much ease as in the vegetable kingdom; and these we can modify or *ameliorate* according to our wants. These races or varieties flourish even when the original animals from whence they sprung have disappeared for ever!

Where is now to be found the original animal to which we owe the *ox,* or the *horse,* or the *camel,* or the *dog?* The original types of these domestic animals have disappeared from the face of the globe. The cow in all probability originated in the animal seen and described by Herberstein (*Rerum Moscovitarum Commentarii,* etc., 1556) in the sixteenth century, under the name of *Thur*. The species to which we owe the horse is extinct; the type of the camel, the original dromedary, the type of the dog tribe are lost for ever.

But they are replaced by numerous varieties of animals so useful to us that they have been called "domestic animals," in producing which man has attended to his own interests.

These modified species of animals are increasing in number daily. The term "domestic" animals should extend over the whole, or, at least, the greater portion of the animal world. Our readers are not accustomed to hear grubs, insects, animalculæ, etc., spoken of as "domestic animals." But do we not rear our *silkworms* with as much care as our *sheep* or our *cows*? Do we not construct houses for our *bees, cochineals, snails, oysters,* etc., as we do for our *rabbits,* our *chickens,* or our *horses*? Are not large fortunes realized by the cultivation of a worm such as the *leech,* or a grub such as the *silkworm,* as readily as by the aid of the *camel* of the desert or the Indian *elephant*? Have we not seen a thimbleful of some new insect or its eggs fetch as high a price in the market as the choicest Cochin-China fowl?

It is too true that these inferior beings are comparatively new to us in this light. But their study affords far greater interest, and, in many cases, undoubtedly more profit, than that of superior animals.

Imagine a man in difficult circumstances endeavouring to gain a livelihood by rearing some new variety of dog, cow, horse, ass, or pig. He would have greater chance of success were he to extract some new colouring matter from the insect world, or discover a means of doubling the produce of the *bee* or the *silkworm*, or a method by which *sponges* and *corals* might be cultivated with as much ease as a *lettuce* or a *cauliflower*.

My endeavour in this volume is to treat of *inferior animals* useful to man, from *insects* downwards to *infusoria* and *sponges*. I leave it to others to write the useful novelties that may concern *Quadrupeds*, *Birds*, *Reptiles*, and *Fishes*. My observations treat of *Invertebrata* only.

Our readers have doubtless heard of a new species of culture which has lately taken a very extensive development. It is called *Pisciculture*, or the breeding of fish, in which many eminent naturalists have met with astonishing success.* Their secret was, however, known long ago to the Chinese. When a

---

* See papers on the subject by Coste, De Quatrefages, and others, and for the artificial propagation of the salmon in Great Britain, see report of a committee, consisting of Sir W. Jardine, Dr. Fleming, and Mr. E. Ashworth, in "Report of British Association," 1856. These researches are facilitated as regards fish by the great fecundity of the latter. Thus, the pike, for instance, produces about 300,000 eggs; the carp, 200,000; and the mackerel, more than *half-a-million*. But this fecundity is still more astonishing in the inferior animals of which we treat here.

Chinaman wished to stock a pool with fish he repaired to some stream where the latter were known to abound, and placed in it bundles of straw, which were soon covered with spawn. After a certain time the straw was withdrawn and placed in his pool, where the eggs were hatched, and the young fish soon became large enough to satisfy their master's appetite.

The writings of Coste, Millet, Géhin, Milne Edwards, De Quatrefages, Remy, and others,* have not only taught us how to stock our streams with magnificent salmon, trout, grayling, etc., but lead us to expect that there will soon exist as many different varieties of trout, salmon, perch, tench, etc., as we have actually of dogs or horses. For certain closely allied species have been crossed so as to produce new varieties or races of fish never before seen.

Similar experiments are being made with inferior animals. The attention of philosophers and practical men is now directed to the latter. We speak now of the *amelioration* of some *insect* species, of the cultivation of a *mollusc* or a *polype*. We begin to see how we can profit by *infusoria* or some other animalculæ.

The following pages will, I trust, give some idea

* Quite recently Mr. Francis and Mr. Buckland have again brought forward the subject of *Pisciculture* in England.

of the extent to which these practical studies are actually pursued; and what animals, a short time since almost ignored, may eventually prove themselves a source of wealth, comfort, and happiness to man.

Chapter II.

Silk-Producing Insects.

The Chemical Nature of Silk—The Spider's Web—Bombic Acid—Detection of Wool in Silk—Great Variety of Insects producing Silk—The Common Silkworm, Bombyx mori—The Golden Tree—The Province of Seres and the Morea—Prolongation of Life in Plants and Animals—Artificial Incubation and Rearing of Bombyx mori—Enormous Appetites—Insects Living without Food—Rate at which the Silkworm Spins—Modes of Destroying the Chrysalis—Calculation basis of Silk-breeding—The two Mulberry Trees—Diseases of Silkworms and their Remedies—Improvement of Bombyx mori—Tussah Silkworms: Bombyx Pernyi and B. Mylitta—Bombyx Cynthia—Extraordinary Qualities of Silk—Other New Species of Silkworm—Spreading of these New Races—The Madagascar Silkworm—Production of Coloured Silk by the Insects themselves—Experiments—Bombyx madrona—Silk of the Clothes-Moth: Tinea—The Paraguay Spider—Ichneumons of the West Indies—Silk Imported into Liverpool.

SILK-PRODUCING INSECTS.

INSECTS are, perhaps, of all animals, those which have proved most useful to man. The silkworm alone, the most important of them all, has been, in a practical point of view, the object of more experiments than any other known creature. Volumes have been written upon it, new varieties are constantly being discovered and reared with hopes of realizing still greater advantages, and, at the same time, investigations are pursued with a view of increasing the produce of the original insect.

The chemical nature of silk, which is secreted through the mouth of the grub from organs resembling the salivary glands of other animals, is very little known. In the body of the silkworm it appears as a viscous liquid, which becomes solid when in contact with the air. If we take a silkworm at the period when he is about to spin his cocoon, and immerse him for twelve hours in vinegar, on opening the reservoir which contains the liquid silk, the latter may be drawn out into threads as thick as a common sized knitting-needle, and of such tenacity that it is impossible to break them with the hands. These thick threads are used to attach hooks to fishing-lines for large fish.

Roard and Mulder have endeavoured to ascertain the chemical nature of silk. The latter chemist has recognized in it a peculiar animal matter, which he terms *fibroin*, or pure silk-fibre. When the liquid silk taken from the body of the grub is placed in acidulated water, it coagulates into a mass of minute white filaments. When secreted by the silkworm a portion of this liquid solidifies and forms a simple thread of silk, which, in contracting, expels from its interior a liquid that solidifies on the surface of the thread, forming a sort of varnish. It is the latter which gives to certain silks their natural yellow colour.

The analysis of Mulder shows that the liquid secretion of silkworms contains about half its weight of pure silk-fibre (fibroin), the remaining portion consists of albumen, two kinds of grease, a species of gelatine, and a slight quantity of a red colouring matter. The spider's web shows a perfectly similar composition.

Mulder has shown that by distilling silk with diluted sulphuric acid, a peculiar product is obtained called *bombic acid*. It may also be obtained by boiling raw silk with water, and evaporating with precaution. This bombic acid is an extremely interesting product, first noticed by Chaussier. It is highly volatile, and possesses a very peculiar strong smell.

It is useful to know how to detect the presence of wool in silken tissues. Lassaigne has given us an easy method of effecting this by showing that a dissolution of oxide of lead in potash will blacken woollen threads, forming sulphide of lead, because wool contains a notable proportion of sulphur. This is not observed with silk threads. If the suspected tissue is coloured, it is necessary to take out the dye before applying the test.

Such are the principal chemical data we possess regarding silk.

This substance is not produced by the silkworm alone; endless varieties of insects, or larvæ of insects, produce it likewise; and we have just seen that the spider's web has a similar composition. Indeed, as we shall see presently, other insects besides the silkworm have been reared with a view of obtaining silk, but as yet only with limited success.

The common silkworm is the larva of a kind of moth (*Bombyx mori*) belonging to the family of Lepidoptera. Much uncertainty has prevailed as to the country in which this *Bombyx* was originally found and reared. It appears evident, however, that the silkworm is a native of China, and that the mulberry tree was cultivated in that country, and known by the name of *The Golden Tree*, two thousand six hundred years before the Christian era.

The insect was afterwards transported to Hindostan, where it was reared successfully for some time in the province of Seres, whence came the denomination *Sericum*, given by the Romans to the product of the silkworm. Persia and many other countries of Asia began in their turns to profit by the cultivation of the *Bombyx mori*, which industry is still carried on there. The vessels and caravans of the Phœnicians carried the Asiatic silk to the principal markets of antiquity.

The mode of producing and manufacturing this precious material was kept secret by many means, and consequently was not known in Europe till long after the Christian era had commenced. It was first learnt, we are told, about the year 550, by two monks, who, having concealed in hollow canes some eggs of the silkworm-moth procured in India, hastened to Constantinople, where the insects speedily multiplied, and were subsequently introduced into Italy, where silk was long a peculiar and stable article of commerce. It was not cultivated in France till the time of Henri IV., who, considering that mulberry trees grew in his kingdom as well as in Italy, resolved to introduce the silkworm, and appears to have succeeded perfectly. However, even in the time of the Emperor Justinian, a certain portion of Greece was covered with such a quantity of mulberry trees (*Morus*), that it received the name of *Morea*, which it retains to the present day.

In an entomological work published in London in 1816, all that is said about the silkworm consists in the following few words: "The most valuable of all moths is the silkworm. . . . . The art of converting its silk into use is said to have been invented in the Island of Cos by a lady named Pamphylis."*

Each female Bombyx lays at least 500 eggs, sometimes this number is much larger. Ten or twelve days afterwards both the male and female moths die. Thus, as soon as they have assured the conservation of the species, they bid adieu to this life. This remark, which is applicable to most insects, is also true for annual plants; and I have shown in another work, that if the coupling of insects be prevented, either accidentally or purposely, it is possible to prolong the period of their existence, and at least to double it. In like manner if an herbaceous plant, such as the mignonette, which dies down at the end of a year, have its

---

* According to Aristotle, the lady's name was Pamphyla, and she must not be mistaken for the woman of the same name who wrote a general history in thirty-three volumes, in Nero's time, and from whose name our English word *pamphlet* is perhaps derived.

flowers carefully cut away as soon as they appear, the plant, instead of remaining an *annual*, continues to live through the winter; and if the same operation be repeated throughout the following year, our mignonette will soon become a *ligneous* vegetable—a real tree; and from that moment the duration of its life is unlimited. This then is the whole secret of the "elixir of life," at least as regards plants and inferior animals. Future research alone can assure us whether the same principle is applicable to higher organisms.

Mr. Spence having remarked that the larva or grub of a certain Aphidivorus fly (a fly feeding upon the *Aphis*, or blight) had lived about twelve months without the slightest particle of food—an example by no means unprecedented in insect life—says : "We can attribute this singular result to no other circumstance than it having been deprived of a sufficient quantity of food to bring it into the pupa state, though provided with enough for the attainment of nearly its full growth as a larva. Possibly the same remote cause might act in this case as operates to prolong the term of existence of annual plants that have been prevented from perfecting their seed; and it would almost seem to favour the hypothesis of some physiologists, who contend that every organized being has a certain portion of irritability originally imparted to it, and that its life will be long or short as this is slowly or rapidly excited."

It is during the spring that the eggs of the silkworm moth undergo artificial incubation or hatching. This is effected by submitting the eggs to a temperature ranging from 16° to 18° Cent.; but sometimes to quicken this operation the heat is raised gradually to 28°. The eggs are hatched in ten or twelve days, when the young larvæ are carefully separated from their former envelopes, and reared on the leaves of the mulberry tree.

M. Perrottet has remarked that silkworms' eggs carried from France to the West Indies, and kept in those hot climates for seven or eight years, could not be hatched until eight or nine months had elapsed, notwithstanding the high temperature, and then only at long and irregular intervals. But when the same eggs were put in an ice-house for four or five months, they were hatched within ten days from their being exposed to the circumambient atmosphere, and nearly all at once.

In establishments where the rearing of silk-worms is carried on upon a large scale, the rooms ought to have a degree of warmth ranging from 16° to 18° Cent., and it is of the utmost importance that the air of these rooms should be perfectly pure. Artificial ventilation is therefore as necessary here as in an hospital. M. Dumas, who has paid great attention to this question, has lately submitted to analysis the air of the rooms in some of the principal silkworm establishments in France, and has found that in many cases this air was devoid of the necessary proportion of oxygen. Indeed, these establishments will never be properly warmed and ventilated unless they adopt the Van Hecke system of ventilation and warming, which is beginning to be generally employed in hospitals, and which is the only system of mechanical ventilation where the heat and the supply of air are completely under control, and can be regulated at will.*

The period which elapses from the birth of the larvæ to the time it begins to spin, varies according to the different climates, and according to the particular species or variety of silkworm cultivated. In China this period is reckoned at twenty-four days; in Italy from thirty to thirty-two days; and in the North of France and Belgium from thirty-two to thirty-four days. In Belgium the culture of the silkworm has begun upon a large scale; there is an extensive establishment in Uccle near Brussels (see Fig. 1), which is the most northern establishment of the kind in Europe.

During this period the grubs change their skin four times, and their appetite, which is enormous, becomes still greater after each moulting. For-

* See "Medical Review," January, 1861, London.

Fig. 1.—Silkworm Establishment at Uccle, Belgium.

tunately they cease eating and become drowsy as the time of moulting approaches. At all times their greatest enemies are damp and noise; they must be kept quiet, and, above all things, clean.

The larvæ born from one ounce of eggs require during their first age, which lasts five days, about 7 lbs. weight of mulberry leaves. After the first moulting, and during the second age, which lasts only four days, they require 21 lbs. of leaf. During the third stage, which lasts a week, they devour 70 lbs. of mulberry leaf; in the fourth stage (also a week), 210 lbs.; and during the fifth stage, from 1200 to 1300 lbs. of leaf. On the sixth day of this last period, they devour as much as 200 lbs. weight of leaf, with a noise resembling the fall of a heavy shower of rain. On the tenth day they cease eating, and are about to undergo their first metamorphosis.*

At this period of their lives they begin to spin their cocoons. The silkworm spins on an average

* Although the *larvæ* or grubs of insects in general are very voracious, as in the example before us, the perfect insect, on the contrary, can live for a long time without food. Thus, Mr. Baker has proved that the beetle called *Blaps mortisaga* can live for three years without food of any kind. The little sugar fish (*Lepisma saccharina*) was shut up in a pill box by Mr. Stephens, in 1831, and found alive in 1833. Leuwenhoek saw a mite which was gummed alive to the point of a needle, live for eleven weeks in that position. Other examples have been noticed in Kirby and Spence's admirable "Introduction to Entomology."

at the rate of six inches per minute. The length of silk furnished by *one* cocoon averages 1526 English feet.

The total quantity of silk spun in one year in Lyons alone amounts to 6,000,000,000,000 of English feet. We cannot be surprised, then, that in the South of Europe the prospect of a deficient crop of silk causes as great a panic as a scanty harvest of grain with us. The average crop is about 80 lbs. weight of cocoons produced from the larvæ hatched from one ounce of eggs. But this harvest is in some cases far greater, and has been known to attain 130 lbs.

In four days the silkworm has completed its cocoon, in which it remains from ten to twenty days in the chrysalis state, from which, in nature, it emerges as a moth. If left to itself the newly formed insect makes its way out of the cocoon by means of a brown liquid it secretes, and which has a corrosive action upon the silk.

To prevent this the chrysalids are destroyed either by placing the cocoons for an hour upon a hot stone, by exposing them for three successive days to the direct rays of the sun, or by heating them by means of vapour in a copper apparatus to $60°$ or $75°$ (Cent.). We are told that the Chinese used formerly to produce the same effect by placing the cocoons in large earthen jars covered with salt, from which they excluded the air.

A certain number of cocoons are put aside to perpetuate the breed. This operation is based upon the calculation that 1 lb. of cocoons are equivalent to one ounce of eggs—that is, that the moths from 1 lb. of cocoons can produce one ounce weight of eggs. We have already seen that one ounce of eggs will produce 80 lbs. of cocoons.

The mulberry tree (a native of China, and of which there are two varieties : the black, *Morus niger*, whose refreshing fruit is well known, and the white, *Morus alba*, upon whose leaves the silkworms breed) is easily cultivated wherever the vine grows, but succeeds very well in more northern climates. I have myself seen the *Morus alba* cultivated with success at Uccle, a pretty spot near Brussels, already alluded to, and I know that experiments of the same kind have met with success in England, Switzerland, Prussia, Hungary, Austria, Russia, etc.

As a consequence of its domestication, the silkworm, though very robust in China, is subject in other countries to various maladies. The worst of these is certainly that called *Muscardine*, which attacks the larva at all periods of its life, but especially while making its cocoon. It is an infectious disease, and can be communicated from the body of a dead larva to that of a living one. This destructive pestilence, for which no efficacious remedy appears to be known, is caused by a parasite fungus, *Botrytis bassiana,* developed in the body of the grub. Absolute cleanliness is the only method by which the invasion of this parasite can be prevented. As soon as it has made its appearance, the sick larvæ must be immediately separated from the others.

Atrophy or *Rachitism* is generally caused by a careless incubation; it is then incurable; but if this disease result from negligence in the breeding, it can be remedied by separating the sick larvæ from the others, and feeding the former upon more delicate leaves. *Gangrene,* which finally reduces the grub to a black fetid liquid, is the result of other morbid affections, and is without remedy. *Jaundice,* which is characterised by a swelling of the skin, which bursts at different parts of the insect's body, is generally caused by sudden atmospheric changes, which trouble the functions of digestion, and is almost always fatal.

In the department of Vaucluse, where, on a small area of land, more than two million of mulberry trees are grown, gangrene, resulting from these and other maladies, is arrested in its course by sprinkling quicklime over the larvæ, by means of a very fine sieve, and then covering them with leaves soaked in wine.

Apoplexy is sometimes determined by sudden changes of the weather, and by bad nourishment;

diarrhœa, dropsy, and some other diseases are generally caused by want of attention on the part of the owners of silkworm establishments.

A remarkable improvement has lately been effected in the breed of the common silkworm (*Bombyx mori*) by M. André Jean, the director of a large silkworm establishment at Neuilly, near Paris, which I had occasion to visit not long ago. This gentleman having communicated his discovery to the *Société d'Encouragement*,* I was invited with M. Dumas to witness the effects of his experiments. A favourable report was afterwards made upon the subject to the Paris Academy. The whole secret consists in causing the largest and finest male and female silkworm moths to breed together. For this purpose M. André Jean places aside for breeding the cocoons which have been spun by the largest caterpillars, and which have a greater weight than the others. From this sorting of the cocoons a very valuable race of silkworms had been created when I visited the establishment, and the inventor is now occupied in distributing the eggs of this new race among the large silk-breeding establishments of France. The larvæ that are developed from these eggs astonish us by their size when compared with the common silkworm.

* See "Bulletin de la Société d'Encouragement," Paris, 1856; and the journal "Cosmos," Paris, 1856.

Up to the present time almost all the silk produced in Europe, and the greater portion of that manufactured in China, has been obtained from the common silkworm (*Bombyx mori*). But new varieties of Bombyx are beginning to be cultivated in Europe, especially in France.

For a very long time considerable quantities of silk have been produced in India from other descriptions of silkworms. Of these the most important are the following :—

First, the *Tussah* and *Arindy* silkworms, whose history has been given with detail by Dr. Roxburgh ("Linnean Transactions," vii. 33). The *tussah* silkworm (*Bombyx Pernyi*) is a native of Bengal, and feeds upon the leaves of the Jujube tree (*Zizyphus jujuba*). Duméril, the celebrated French naturalist, cultivated it for some time, as an experiment, upon the leaves of another tree, *Jambosia pedonculata*, and M. Guérin Menneville has bred this *tussah* worm exclusively upon *oak leaves*. Besides which, it is known to live upon a plant called *Terminalia alata glabra*. So that this grub has the advantage of being what is termed *Polyphytophagous*, that is, it can be made to feed upon different kinds of leaves. This fact has been observed with some other species of Bombyx. It is certainly a great advantage to those who undertake to introduce it into Europe.

The silk of the *tussah* worm is much coarser

than that of the common silkworm, and of a darker colour. With it are clothed one hundred and twenty millions of Chinese, Brahmins, etc., and it would doubtless be useful to the inhabitants of the New World and the South of Europe, where a light, cool and at the same time cheap and durable dress is much wanted. Garments made of tussah silk will wear, when in constant use, for ten or twelve years.

Tussah silk is also produced by another species of Asiatic moth, *Bombyx Mylitta*, which has lately been successfully reared in France by M. Guérin Menneville, at Paris, and also at Lausanne. Its leather-like cocoons are composed of silk so strong that a single fibre will support, without breaking, a weight of one hundred and ninety-eight grains. It also feeds upon a great variety of leaves, among others upon oak leaves. The eggs of this moth have been known to hatch in Siberia before the appearance of leaves upon the oak tree. The only way of preventing the larvæ from starving in such cases, is to cut branches from the oak and place them in vessels of water. The leaves are thus made to shoot out quickly, and the grubs are fed upon them until the oak tree is covered with foliage. The natural enemies of these larvæ are birds, bats, ants, some species of frog, serpents, and foxes, who enjoy them exceedingly.*

* The fox will also eat beetles, and attack bees' nests for honey.

The *Arindy* silkworm (*Bombyx Cynthia*), discovered in Bengal, feeds upon the castor-oil plant (*Ricinus communis*). This curious plant, which in India and Africa is a large tree, becomes in our climate a small herbaceous annual. The silk produced by *B. Cynthia* is remarkably soft and glossy; it cannot be wound off the cocoon, and is therefore woven into a kind of coarse white cloth of a loose texture, used for clothing, and for packing expensive fabrics. Its durability is so great that a man's lifetime is insufficient to wear out a garment made of it.

M. Guérin Menneville, who has experimented with this silkworm, informs us that the transformation of its chrysalis into a moth may be artificially suspended for a period of seven months. The chrysalis of our common silkworm may be kept in this state for a period of two years if the temperature be cool. If the latter rises from 15° to 18° Cent., the moth comes forth in eighteen or twenty days; but it is a *general rule* with insects that the time they remain in the chrysalis state depends upon the temperature.

The way in which many insects resist cold is truly wonderful. Many larvæ and chrysalids may be frozen until they become as brittle as glass, and after having remained for some time in this state, they revive by the application of warmth. Spallanzani once exposed the eggs of the silkworm to an intense degree of cold, produced by an artificial freezing mixture, in which they remained for five hours without being frozen, the thermometer of Fahrenheit having fallen to 56° below zero, although the liquid portions of an insect's egg has been shown by John Hunter to freeze at 15° Fahr. The eggs submitted by Spallanzani to this treatment were afterwards hatched.

In 1854 the Governor of Malta made several reports upon the *Bombyx Cynthia* for the information of the Society of Arts. It had been introduced into Malta from India that year, and appeared hardy and wonderfully prolific. Yet it failed in 1855. The author of these observations had, however, previously distributed its eggs throughout Italy, France, and Algeria, and, continuing to watch the trials made in these countries, he found that the new silkworm had flourished and had been carried into Spain and Portugal. He therefore reintroduced it into Malta. At the end of July 1857, he received a few eggs by post in a quill from Paris, and these have multiplied in an extraordinary manner. The winter season (December) appeared to affect the caterpillars even in Malta—they grew slower than in summer, but nevertheless appeared healthy.

In France experiments are being made on the silk of the *B. Cynthia*, which is found to be very fine, and to take dyes admirably. The cocoons are carded and afterwards spun. It has been discovered that the chrysalis in extricating itself from the cocoon does not cut the thread as had been asserted, and the French have partially succeeded in unwinding the cocoons after the exit of the moth.

The natural climate of *B. Cynthia* lies upon the borders of the tropics, hence the difficulty experienced in keeping the insect during the winter in European climates. It is spreading, however, rapidly over the globe. The Governor of Malta sent it to the West Indies in 1854. The French have forwarded it to the Brazils, to the Southern States of North America, and to Egypt. It has likewise spread from Malta to Sicily, and 127,000 cocoons have recently been sent from Algeria to be manufactured in Alsace. Although its natural food is the castor-oil plant, it will live and thrive, we are told, upon the Fuller's teasel (*Dipsacus fullonum*).

Besides these varieties of silkworm, the members of the Société d'Acclimatization of Paris are about to make experiments with other species, such as *Bombyx Bauhinia*, *B. Polyphême*, *B. Aurota*, etc., all exotic insects, at present little known.

In Victoria, according to the "Australian and New Zealand Gazette," of 1858, a native variety of silkworm has been discovered in the bush. Mr.

Whyte has forwarded cocoons to several establishments. The product of this new insect is said to be of a very superior kind; and the insect is extremely abundant in that colony.

It is not very long since that the famous Madagascar silkworm created much sensation in Europe, and hopes were entertained of rearing it in France. The most remarkable peculiarity of this insect is that several of its larvæ spin together and produce a cocoon as large as an ostrich egg.

Some experimenters have endeavoured to make the silkworm produce silk ready dyed. On this point we know that when certain colouring matters extracted from the vegetable kingdom are mixed with the food of animals they are absorbed without decomposition and colour the bones and tissues of the body. Starting from this fact, Messrs. Barri and Alessandrini, in Italy, sprinkled certain organic colouring matters over the mulberry-leaves on which the silkworms were feeding. M. Roulin, in France, employed in the same way the colouring matter known as *chica*. These attempts have met with partial success only, up to the present time; but they deserve to be continued. Coloured cocoons were thus produced several times. Some observers assert, however, that the silk was not really secreted in a coloured state, but that the colouring matter sprinkled on the leaves merely adhered to the body of the grub, and coloured the cocoon mechanically during its construction. This appears to be the reason why the coloured silk that was obtained in these experiments was neither uniform in tint nor of a good colour. Others, however, still persist in a contrary opinion. M. Roulin commenced his experiments by sprinkling *indigo* over the mulberry-leaves, and obtained *blue cocoons;* he then experimented with *chica,* a fine red dye extracted from the *Bignonia chica,* which the Indians of Oronoco employ to dye their skin, and obtained cocoons of a red colour, with a tolerably uniform tint, and of a permanent dye. He still continues these investigations, hoping to obtain silk ready dyed of all kinds of colours.

Whatever may be thought of these experiments as they now stand, they are novel, and should therefore be encouraged. It would, probably, be worth while to try the effect of the famous new green dye, *Lo-kao,* mixed with the diet of the silkworm. This colour, which is one of the most beautiful and most extraordinary dyes ever yet produced, has great affinity for silk; it is extracted from several species of *Rhamnus,* and we have seen that certain varieties of silkworm feed upon the leaves of plants (*Zizyphus,* etc.) of the same family.

Kirby and Spence have informed us that Don Louis Née observed on *Psidium pomiferum* and *P.*

pyriferum ovate nests of caterpillars eight inches long formed of grey silk, which the inhabitants of Chilpancingo, Tixtala, etc., in America, manufacture into stockings and handkerchiefs. Great numbers of similar nests of a dense tissue were observed by Humboldt in the provinces of Mechoacan and the mountains of Santa Rosa, at a height of 10,500 feet above the level of the sea, upon the *Arbutus madróna* and other trees. The silk of these nests is produced by the larvæ of *Bombyx madróna*, who live in society and spin together. It was an object of commerce with the ancient Mexicans, who made it into paper. Handkerchiefs are still manufactured of it in Oaxaca.

It is a doubtful question whether the breeding of any European moths will ever become a source of advantage. Experiments have already been made on certain varieties of clothes-moths (*Tinea*). Mr. Habenstreet, of Munich, experimented some years ago upon a species called *Tinea punctata*, or *Tinea padilla* (Fig. 2), closely allied to *T. Evonymella;* the larvæ of the former were made to spin upon a paper model suspended from the ceiling of a room. To

Fig. 2.—Tinea padilla (Silk-spinning gnat).

this model, any form or dimensions could be given at will, the motions of the larvæ being regulated by means of oil applied to those parts of it which were not intended to be covered. The investigations showed that on an average two of these larvæ can produce a square inch of silk, and when employed in great numbers their produce is astonishing. Mr. Habenstreet succeeded thus in manufacturing an air-balloon about four feet in height, one or two shawls, and a complete dress with sleeves, without any seams. The tissue thus curiously produced resembled the lightest gauze, which it surpassed in fineness. We are told that the Queen of Bavaria once wore a robe of this description over her court dress.

On mentioning these experiments to my friend, M. Babinet, of the French Academy of Sciences, he said the only thing that could be urged against the use of this silk of the *Tinea punctata* was its excessive lightness; the slightest breath of wind is sufficient to carry away a whole dress. We will only add to what we have already said concerning these silk-producing insects, that De Azora speaks of a peculiar spider in Paraguay which envelopes its eggs in a yellow cocoon of an inch in diameter, and whose silk is spun into dresses by the inhabitants of Paraguay. The colour of this silk is very permanent.

The Ichneumon flies of the West Indies, which feed upon the indigo and cassada plants, furnish a silk of peculiar whiteness, which is not yet employed.

Silk of *Bombyx mori* is imported in the raw state,

as spun by the insect, into Liverpool, at the rate of about 57,000 lbs. annually. Tussah silk from *B. Mylitta* arrives in Liverpool from the East Indies in quantities which vary from 2000 lbs. to 12,000 lbs. per annum.

Chapter III.

Colour-producing Insects.

The Kermes—Latreille and his genus Coccus—Coccus ilicis—Crimson of the Romans—Brussels and Flemish Tapestries—Coccus polonicus—Coccus of the Poterium—Coccus uva-ursi—The Cochineal—Coccus cacti—Plants on which the Cochineal lives—Nopaleries—Grana sylvestra and Grana fina—Rearing of Cochineal—The Cochineal at Teneriffe—The Bluebottle Fly and the Aphides—Generation extraordinary—Two new Cochineals in Australia.—Coccus fabæ (or Aphis fabæ) in France—Its peculiar ColouringMatter—Lac—Carminium, its discovery and properties—The Colouring Matter of the Cochineal discovered in the Vegetable World—Carmine—Influence of Light in the Manufacture of Colours—Rouge for the face—Ink—The Cynips—Caprification—Diœcious Plants—Ripening of Figs in the East—Gall-nuts—Cynips-gallæ-tinctoriæ—Theory of the Formation of Vegetable Tumours—Analysis of Gall-nuts—Their products and Uses—Cynips quercus folii—On the Formation of Grease by Animals—Other Insects producing Dyes—Aphis pini—"Money-spiders"—The Magenta Dye and Cochineal.

COLOUR-PRODUCING INSECTS.

COLOUR-PRODUCING insects come next, perhaps, in importance to those we have already noticed. The cultivation or breeding of these useful little animals forms one of the most interesting and profitable branches of industry.

I shall begin by speaking of the *Cochineal*, which will constitute the most important feature of this chapter; but I prefer drawing attention, in the first instance, to the Kermes (or Chermes), a little insect of the same genus as the former, known and employed long before the cochineal insect was discovered.

The insects of which I am about to treat all belong to Latreille's genus *Coccus*, in the family of the Hemiptera. The number of species belonging to this genus being very great, and being possessed of extraordinary colouring properties, they constitute a wide field for research and experiment. The more so, as very few are, as yet, cultivated to any extent, although many species appear to possess all the necessary qualifications, and many others are ignored in a practical point of view.

The Kermes (*Coccus ilicis*, Latr.) has been employed to impart a scarlet colour to cloth from the earliest ages. It was known to the Phœnicians under the name of *Tola*, to the Greeks as *Kokkos*, and to the Arabians and Persians as Kermes or Alkermes (*Al* signifying *the*, as in the Arabian words alkali, alchymy, etc.). In the Middle Ages it received the epithet *Vermiculatum*, or " little worm," from it having been supposed that the insect was produced from a worm. From these denominations have sprung the Latin *coccineus*, the French *cramoisi* and *vermeil*, and our *crimson* and *vermillion*.

The *Coccus ilicis*, or Kermes, is found in many parts of Asia, the southern countries of Europe, and the south of France, where it is very common. The first person who made mention of this insect appears to have been Pierre de Quiqueran, who spoke of it as early as 1550. Its history was afterwards written by Nissole in a paper addressed to the Paris Academy of Sciences in 1714, and by Reaumur in the tome iv. of his "Mémoires pour servir à l'Histoire des Insectes." The females resemble a pea in size and form, whence they have been frequently taken for seeds. The insect lives upon a small evergreen oak, the *Quercus coccifera*, L., and yields a brownish red colour, which alum turns to a blood-red tint.

Dr. Bancroft has shown that when a solution of

tin is used with kermes dye, as with cochineal, the kermes is capable of imparting a scarlet quite as brilliant as that produced by the cochineal itself, and to all appearance more permanent. But on the other hand we know that one pound of cochineal contains as much colouring matter as ten or twelve pounds of kermes. However, we are told that it was with the latter insect that the Greeks and Romans produced their crimson, and from the same source were derived the imperishable reds of the Brussels and other Flemish tapestries. Cochineal has supplanted kermes, and the latter is now only cultivated by some of the poorer inhabitants of the countries in which it abounds, more particularly in India and Persia, and by the peasantry of southern Europe.

Another species of kermes, the *Coccus polonicus*, Latr., sometimes known as the *scarlet grain of Poland*, is very common in Poland and Russia. Before the introduction of cochineal this insect formed a considerable branch of commerce. In the neighbourhood of Paris, and in many parts of England the *C. polonicus* is found upon the roots of *Scleranthus perennis* (perennial knarvel), a plant that is not uncommon in Norfolk and Suffolk. The colour which it furnishes is nearly as beautiful as that of the cochineal, and capable of giving the same variety of tints. The insect was formerly collected in large quantities for dyeing red in the Ukraine, Lithuana, etc., and though still employed by the Turks and Armenians for dyeing wool, silk, and hair, but more particularly for staining the nails of the Turkish women, it is rarely used in Europe except by the Polish peasantry.

The same may be said of other species which the cochineal has completely eclipsed, such as the *Coccus* found upon the roots of *Poterium sanguisorba*, an insect formerly used by the Moors for dyeing silk and wool a rose colour; and the *Coccus uva-ursi*, which, with alum, dyes crimson. All these species owe their colouring properties to a principle called *carmine*, which I shall refer to presently.

The discovery of the cochineal has not prevented experiments being daily made with these and other species of Coccus, which we shall mention hereafter.

The cochineal (*Coccus cacti*, Latr., Fig. 3) was already in use in Mexico when the Spaniards arrived there in 1518; its true nature was not, however, ascertained till upwards of a century later. Although Acosta declared cochineal to be an insect as early as 1530, it required the labours of many naturalists from that period till 1714, to place its real nature beyond doubt, so generally was it supposed to be the seed of a plant.

Fig. 3—Coccus cacti, Latr. (Cochineal magnified).

The *Coccus cacti* is a native of Mexico, where it lives upon different species of *Cactus* or *Opuntia*. The plants chiefly cultivated in hot climates for breading cochineal are the *Cactus coccinellifer*, *C. opuntia*, *C. tuna*, *C. paresxia*, etc. The first of these is also called *Opuntia coccinellifera*, and is known as the *Nopal*, although it appears, from Humboldt's account, that these plants are two distinct species, the latter being probably the *Cactus opuntia* of Linnæus. However, the insect thrives equally well on both.

The cochineal, which comes to us in the form of a small shrivelled grain of a reddish colour, covered with a sort of white down, was formerly only cultivated in Mexico. The female alone is of any commercial value. The male enjoys only a short life, and generally dies at the age of one month; its wings are as white as snow. The females fix themselves firmly by means of their proboscis on to the plant which serves them as a habitation, and never quit this spot. Here they couple with the male insects, and increase considerably in size. Each female lays several thousand eggs, which proceed through an aperture placed at the extremity of the abdomen, and pass under the body of the mother-insect to be hatched. The mother-insect then dies, and her body dries up and forms a kind of shell or envelope in which the eggs are hatched, and from whence the little cochineals soon proceed.

The cultivation of the Nopal and its cochineal was originally confined to the district of La Misteca, in the State of Oaxaca, in Mexico, where some plantations contain upwards of 60,000 separate plants set in straight lines, each being about four feet high, which height it is not allowed to exceed, so that the insect may be easily gathered. The flower is always carefully cut away. These plantations are called Nopaleries (*Nopaleros*), from the name of the plant, which is chiefly cultivated for cochineal in Mexico. We are told that the greatest quantity of this insect employed in commerce is produced from small nopaleries belonging to Indians of extreme poverty.

Two varieties of cochineal are gathered and sent into the market, the wild kind from the woods, called by the Spaniards *grana sylvestra*, and the cultivated, or *grana fina*. The former is decidedly inferior in quality to the latter, and furnishes far less colouring matter.

The insect in its natural state is of a dark-brown colour, but fine cochineal when well dried and properly preserved should have a grey tint bordering on purple. The grey colour is owing to the downy hair which naturally covers its body, and to a slight quantity of wax. The purple shade arises from the colouring-matter extracted by the water in which the insects have been killed.

The wild variety (*grana sylvestra*) loses by cultivation a good deal of its cottony or downy appearance, and doubles in size; it is then known as *grana fina*.

Real cochineal is detected by the following character:—it is wrinkled, with parallel furrows across the back of the insect, which are intersected in the middle by a longitudinal furrow. This serves to distinguish the true cochineal from any fictitious preparation. Sometimes smooth black grains called "East India cochineal" are mixed with the genuine article, but an experienced eye easily detects the fraud.

A French naturalist, Thieri de Menonville, exposed himself to great dangers for the sake of observing and studying the cultivation of the cochineal in Mexico, in order to enrich by its means the colony of St. Domingo. He carried there the two varieties mentioned above, along with the nopals on which they lived. He discovered also the variety *sylvestra* living upon the *Cactus paresxia*, at St. Domingo—a discovery that was not without value to Bruley—and soon set about the rearing of this interesting little insect; but death cut him short in his experiments, and Bruley continued them with much success. The posthumous work of Thieri was afterwards published, and may be consulted with profit by rearers of cochineal to this day.*

It was generally thought for a long time, and, indeed, it is still believed by many, that the cochineal derives its colour from the nopal on which it lives, the flowers of which are red, but Thieri observed that the juice on which the insect nourishes itself is of a green colour, and, moreover, that the cochineal can be reared and multiplied upon certain species of opuntia, whose flowers are not red. I should mention here, however, that in the "Philosophical Transactions," vol. 50, it is stated that when *Cactus opuntia* is given to children, their urine becomes of a lively red colour, and we shall see presently that *carminium*, the colouring-matter of cochineal, has been discovered in the vegetable world, in a plant of the Orchidæ family.

The wild cochineal has been found in many parts of North America. Dr. Garden observed it in South Carolina and Georgia; it has since been discovered in Jamaica and Brazil. Anderson thought he had seen it wild in Madras, but the species he took for the true cochineal turned out to be another species of Coccus, a kind of Kermes.

---

* "Traité de la Culture du Nopal et de l'Éducation de la Cochenille dans les Colonies Francaises de l'Amérique, précédé d'un Voyage à Guaxaca." Par M. Thieri de Menonville. "Annales de Chimie," tom. v.

When preserved in a dry place, cochineal retains its colour for an unlimited time. Hellot made experiments with some dried cochineal that had been kept a hundred and thirty years, and found their colour as vivid as that furnished by the insects just taken from the *Cactus*.

The poor Indians spoken of above establish their nopal plantations on cleared ground, on the slopes of mountains or ravines, two or three leagues from their villages, and when properly cleaned, the plants are in a condition to maintain the insects for three years. In spring, the proprietor of a plantation purchases as stock a few branches of *Cactus tuna*, laden with small cochineals recently hatched, called *semilla* (seeds). The branches may be bought for about three francs the hundred; they are kept for twenty days in the interior of the huts, and are then exposed to the open air under a shed, where, owing to their succulency, they continue to live for several months. In August and September the female insects big with young are gathered and strewn upon the nopals to breed. In about four months the first gathering, yielding twelve for one, may be made, which, in the course of the year, is succeeded by two more profitable harvests. In colder climates the young insects (*semilla*) are not placed upon the nopals until October or even December, when it is necessary to shelter them with rush mats, and the harvest is proportionately later. Much care is required in the tedious operation of gathering the cochineal from the cactus or nopal; it is performed with a squirrel's tail by the Indian women, who for this purpose squat down for hours together beside one plant. The insects are killed either by throwing them into boiling water, by exposing them in heaps to the sun, or by placing them in ovens. Seventy thousand dried insects weigh on an average one pound. Dr. Bancroft estimated the consumption of cochineal in England at one hundred and fifty thousand pounds per annum, worth about £375,000 sterling, and when Alex. Von Humboldt wrote his "Political Essay on New Spain," the quantity of cochineal exported from Mexico was worth upwards of £500,000 per annum. Since that period the cultivated or "domestic" cochineal and the cactus on which it feeds have been introduced into Spain, India, and Algiers, etc., where its cultivation has greatly increased.

Professor Piazzi Smyth has given an account of the introduction of the cochineal into Teneriffe: "Who would have thought in 1835," says he, in the account of his astronomical observations in that island, "that the years of the grape-vine of Teneriffe were numbered?"

Teneriffe had effectively been a vine-producing country for three hundred years; and when a gentleman introduced the cactus and cochineal there from Honduras, he was looked upon as an eccentric man, and his plantations were frequently destroyed at night. However, when the grape disease broke out, Orotava was gradually forsaken by vessels in quest of wine which could no longer be supplied; and with starvation staring them in the face, the inhabitants turned to cochineal growing: wherever a cactus was seen upon the island, a little bag of cochineal insect was immediately pinned to it. The essay succeeded admirably. An acre of the driest land planted with cactus was found to yield three hundred pounds of cochineal, and, under favourable circumstances, five hundred pounds, worth £75 to the grower. Such a profitable investment of land was never before made. In the south of Teneriffe, the cochineal insect thrives best, and two harvests are made in the year; in the north of the island only one harvest is made, and the growers are consequently obliged to buy fresh insects every season from the south, as the little beings cannot survive the northern winter.

Now, we know from experiments that in warm climates as many as *six harvests* of cochineal may be made in the year; and these are so abundant, the first more especially, that more than one million pounds weight of cochineal arrives in Europe every year. The cactus knows no greater enemy than rain;

it is, therefore, essential to protect it from the wet.

The cochineal grower must also scrupulously avoid the mixing of different species of *Coccus* on the plants; even the wild variety (*sylvestra*) must be kept away from the cultivated (*fina*), or the latter will become thin and maladive, and breed a cross variety, which is inferior in quality. After gathering the insects, the plants must be washed with a sponge before being strewn with the mother-insects. In 1853 there were already seventeen French *nopaleries* in Algiers; at which epoch M. Boyer collected there 2000 francs worth of cochineal from three thousand nopals, which occupied only one-sixteenth of an *hectare* of ground.

The *Coccus cacti* or cochineal from Mexico is imported occasionally from South America to Liverpool: in 1855 one hundred and seventy-three hundred weight arrived.

Like the "Blue-bottle fly" and the *Aphides* (or blight), the cochineal insects (*Coccus*) do not always lay eggs like other insects, but give birth to young *larvæ*, having very close resemblance to their mothers. Thus, with *Aphides* and *Coccus*, we observe the following curious phenomena:—In the early part of the year the female insects do not lay eggs, but bring forth young insects (without previous fecundation), the whole of which are also females. These bear young again, without the concourse of any male insect, and so on for about nine generations. Finally, in autumn, the last generation of females give birth to insects of both sexes. The sexes unite, the males die, and the females *deposit eggs* upon the branches and die also. These eggs pass the winter season on the spot, and in the spring give birth to females which reproduce similar females, and so on throughout the year without the concourse of the other sex. This is certainly one of the most extraordinary phenomena Natural History has revealed to us. In speaking further on of the genus *Melo*, I shall refer to similar curiosities in the embryo life of insects, and when speaking of *Infusoria*, I will make known some extraordinary facts lately discovered, with regard to their development also.

When Leuwenhoek first announced that the aphides were viviparous, and that he suspected they were born without previous fecundation, the researches of naturalists were immediately directed to this point. Reaumur showed that aphides were, indeed, viviparous; he then tried to rear them in perfect solitude, but his insects died, and his experiment failed. It was reserved for Bonnet to confirm the ideas of Leuwenhoek. Bonnet reared aphides in complete solitude from the time of their birth, and in a few days remarked that they brought forth young. He immediately placed the latter in confinement, and observed them give birth to other young aphides. By following up the experiments he saw produced before his eyes *nine generations* of aphides, successively born without the concourse of the two sexes. But it had been certainly ascertained that there exist male and female aphides, and it was also given to Bonnet to observe their *accouplement*. In autumn he saw the little winged aphides couple with the females, which are much larger, after which he saw no more young aphides appear: the females laid eggs, which both Bonnet and Reaumur looked upon as avorted fœti, as they never seemed to hatch. Lyonnet was more fortunate: he observed the hatching of eggs laid by the *aphis* of the oak-tree. Dutrochet, in a short paper read in 1818, at the Paris Academy of Sciences, shows the complete organization of the generative organs of the male and female aphides, and has come to the conclusion that these insects are not *hermaphrodite*, as Reaumur supposed, but that the opinion professed by Trembley, that the fecundation which takes place in autumn is sufficient to render fertile the nine successive generations of females, appears most probable.*

The marvellous tinctorial properties of the cochi-

* Dutrochet's paper was subsequently published in 1833 (' Ann. des Sciences Naturelles," vol. xxx.)

neal insect renders interesting the discovery lately made of two new species of cochineal, both natives of Australia, which have not yet been described by entomologists. They were discovered by Mr. Child. One of them lives upon a species of *Mimosa*, the other on a species of *Eucalyptus*. They produce four or five generations during the year. A short time ago M. Guérin Menneville presented to the Paris Academy a new *indigenous* cochineal which was found living upon some weeds of our own climate, and from which a magnificent scarlet dye can be obtained. This new insect has been denominated *Coccus fabæ*, as it may be successfully reared upon the bean, on the stalks of which vegetable it appears to have been originally discovered. It was afterwards found upon the sainfoin.

Coccus fabæ was discovered by M. Guérin Menneville in the South of France. The discovery terrified him not a little, for should *Coccus fabæ* multiply under favourable circumstances as rapidly as these kind of insects usually do, it would become a disastrous source of *blight* to beans and sainfoin, and possibly to other plants. He then thought of turning his discovery to account, and proclaimed his new insect an extremely useful one, that by proper cultivation might one day replace the exotic cochineal. M. Chevreul, who examined the colouring matter it produced, pronounced it to be a peculiar scarlet, which, until then, could only be obtained by artificial mixtures. It appeared to have a decided advantage over real cochineal as regards the dyeing of wool, but only if the new insect could be procured at a cheaper rate than cochineal, as it was much less rich in colouring matter than the latter. Moreover, the colouring matter of this new insect is not *carminium*, but a perfectly distinct substance. Now all insects belonging to the genus *Coccus* yield carminium, therefore M. Guérin's new insect is certainly not a *Coccus*, but probably, as M. Duméril stated, an *Aphis*, whence *Aphis fabœ* would be its proper name.

A new dye, called Canadian cochineal, has been lately prepared by Professor Lawson, of Queen's College, Canada, from an apparently new species of *Coccus*, which was noticed in the summer of 1860, on the common black spruce (*Abies nigra*) in the neighbourhood of Kingston. The new dye is very similar to cochineal, but, unlike it, can be produced in temperate climates.

I must here briefly notice the little insects which furnish *lac*, and which belong to other species of *Coccus*. Lac is a dark red substance which was supposed to be formed by *Coccus lacca* (or *Coccus ficus*) as bees form their cells. But from the analysis of this substance made by Unverdorben, it appears to consist of five sorts of resins mixed with a little wax, colouring matter, and grease, and that it exudes from the branches of *Zizyphus jujuba* and other trees, after they have been pricked by the *Coccus lacca*. It is collected from various trees and shrubs in India, where it is very abundant, and has the appearance of a concrete juice adhering to and encircling the branches. Chevreul discovered that its red colour was owed to *carminium*—the principle of the cochineal, and therefore its colour is certainly produced by the insect *Coccus*.

There are several varieties of this substance, known in commerce as stick-lac, seed-lac, and shell-lac. *Stick-lac*, when it is in its natural state, adhering to the branches (Fig. 4); *seed-lac* when separated, pulverized, and the greater portion of colouring matter extracted by water; *lump-lac*, when melted and made into cakes; *shell-lac*, when strained and formed into transparent plates.

Two other products are also brought from India. They are chemical preparations for dyeing, called *lac-lake* and *lac-dye*.

Fig. 4.
Stick-lac.

In the latter country lac is used to manufacture beads, rings, and other ornaments. Mixed with sand, it is used to construct grindstones. In this country it is used principally for varnishes, japanned ware, and sealing-wax, and sometimes as a substitute for cochineal in dyeing scarlet. Formerly large quantities of lac-lake precipitated from an alkaline solution of the resin by alum, was manufactured in Calcutta and exported to England. At present it is imported from the East Indies in two forms. Shell, stick, and seed-lac (the resinous exudation) arrives in Liverpool at the rate of about two hundred tons per annum. It is principally used for varnish. Lac-dye or cake-lac, and lac-lake (the colouring matter of the insect combined with alumina, etc.) arrives in Liverpool at the rate of about seventy tons per annum. It is used exclusively for dyeing.

Carminium, the colouring matter of the cochineal, is a very interesting substance. It was first extracted from the *Coccus cacti* by Pelletier and Caventou in 1818. They observed that it formed with alumina a magnificent lake, which they called *carmine.* This lake was, however, previously formed many years before by Dr. Bancroft. M. Lassaigne discovered carminium in the kermes (*Coccus ilicis*), and Chevreul asserted that it existed also in lac-dye (product of the *Coccus lacca*). It has also been extracted from *Coccus polonicus,* etc.

The reason why all these insects cannot be employed so advantageously as *Coccus cacti,* is simply because they yield a much smaller proportion of carminium, and contain a greater quantity of grease, etc. This is so true that if the greasy matter be previously separated by pressure from *Coccus polonicus*, this insect can be employed weight for weight with the same advantage as the genuine cochineal.

Carminium may be obtained by treating pulverized cochineal, first by ether to extract the greasy matter, and then by alcohol. The product thus obtained is treated once more in the same manner, when, by evaporation of the alcohol, carminium is deposited as a granular substance of a red-purple colour. If carminium be combined with oxide of lead, we obtain a violet compound, which, when decomposed with sulphuretted hydrogen, yields a transparent colourless liquid, by the evaporation of which a new substance is deposited in colourless crystals. These absorb oxygen from the air and become carminium.

In August 1856, M. Belhomme made the beautiful discovery of carminium in the vegetable kingdom; he found it in the petals of a plant of the Orchidæ family, the *Monarda didyma*, L. This plant, which has been known to horticulturists for some time, is a native of North America. When its petals are placed in water, they yield to the liquid a crimson colouring matter in every respect similar to the carminium of the cochineal. Some time ago the author of this work thought he had discovered carminium in the bark of the alder tree, but it turned out to be another colouring matter, still more interesting in a chemical point of view.

The colouring matter of the cochineal, like that of madder, or Turkey-red, becomes yellow by the action of acids, but we can distinguish it from the latter, for when carminium is separated from the acid, it appears with its usual red colour, whilst madder remains yellow.

Light has a peculiar action upon carmine—the beautiful crimson lake obtained by precipitating an alkaline solution of cochineal by alum. Mr. Hunt has shown that when this lake is prepared in the dark, it is of far less brilliant a colour than when prepared in the sunshine. The same fact has been observed for other colours, such as Prussian blue, etc.

The colouring matter for the face called *rouge*, employed upon the stage—and sometimes off it—is made by mixing half a pound of prepared chalk with two ounces of freshly prepared carmine. This is the only red colouring matter that should be tolerated for this purpose, as it is perfectly harmless; the other products sometimes sold as such are extremely hurtful, from their venomous properties. M. Chevalier of Paris has very recently made a long report upon the sufferings produced among actors and actresses in Paris by the use of poisonous colours containing lead, mercury, arsenic, and other toxic principles.

*   *   *   *

I shall now turn to gall insects, or *Cynips*, to which we owe many useful products.

If *ink* were the sole product of the insects which produce the gall-nut, we should not be so much indebted to them, as ink can be produced in a variety of manners. But we shall see that the *Cynips* furnish us with other substances useful to mankind. Although the insect which produced the gall-nuts found in commerce was not known to Linnæus or to Fabricius, it belongs to their genus *Cynips*—a genus composed of small four-winged flies, and classed in the family of *Hymenoptera*. Some of these flies are remarkably useful to the Greeks in their process of *caprification*. A diœcious fig-tree, very common in the East, would indeed be comparatively useless but for their aid. By a diœcious plant is meant one in which the male and female flowers are found on different individuals. In most plants the two sexes are united in the same flower, but in others, such as the hop, the nettle, some willows and figs, etc., the male flowers (stamens) are found on one individual, the female flowers (pistils) on the other. Now, as no fruit can ripen without the concourse of these two kinds of flowers,* the female fig-trees of the East are apt to become sterile when removed from the immediate vicinity of the male plants. On the other hand a certain species of *Cynips* is known to abound in the flowers of the latter; so that to render their female trees fertile, the Greeks imagined the process of *caprification,* which consists in this : As soon as the male flowers are in full bloom, they are cut off and strung into garlands, which are hung upon the branches of the female trees. The *Cynips* in their passage from the male to the female flowers, carry with them the pollen of the former, and so the conditions of fertility are ensured.

There are many descriptions of gall-nuts, but those which are mostly esteemed for industrial purposes are the gall-nuts of the East, exported chiefly from Aleppo, Smyrna, etc. They are the product of an insect first described by Olivier, and now generally known as *Cynips gallæ tinctoriæ.*

When an insect of the *Cynips* kind is about to lay its eggs, it makes a slight incision in the leaves of certain plants into which it deposits its eggs. The sap of the plant thus wounded flows rapidly to this spot—a separate incision is made for each egg —and in course of time a small excrescence is formed. The eggs hatch and the new-born *larvæ*

* There are two apparent exceptions to this rule, namely, the *Cælobogyne*, or batchelor plant, and *Hemp*.

nourish themselves on the tissue of the excrescences, thereby causing the sap to flow again to these parts. As the little ball or wart grows in size, its interior is excavated more and more by the increasing appetite of the larvæ, until the sides of the excrescence have become tolerably thin. The larva thus becomes a chrysalis, and when its metamorphosis is completed, the perfect insect without much difficulty bores through the gall-nut and makes its exit.

There are galls of all sorts and sizes, many of which possess very curious forms; but each different variety is produced by a distinct species of *Cynips*.

Reaumur and Malpighi, to whom we owe our knowledge of the formation of gall-nuts, assure us that one of these, however large, attains its full size in a day or two, and that those which spring from leaves constantly take their origin from the nerves or veins of the leaf.

The galls produced by *Cynips gallæ tinctoriæ*, fetch a high price in the markets. They were formerly analysed by Sir Humphry Davy, who found in them 63 parts of cellulose or vegetable fibre, 26 of tannic acid, 6·2 of impure gallic acid, 2·4 of mucilage, and 2·4 of ash or mineral matter. To the tannic acid they owe their highly astringent property, on account of which they are employed in medicine—their gallic acid is indispensable for photography: by the action of heat it is converted into pyrogallic acid, which is still more useful to photographers. By mixture with salts of iron they produce ink and black dyes, and tincture of galls is a reagent constantly employed in chemical laboratories.

These gall-nuts are found upon the leaves of an oak tree (*Quercus infectoria*, L.) The little red oak balls found in our oak leaves are owed to the *Cynips quercus folii* (Fig. 5); they also can be em-

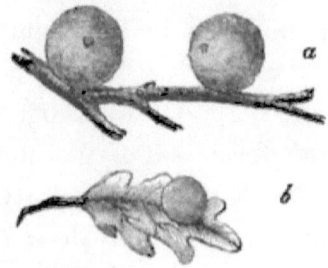

Fig. 5.—*a*, Foreign galls; *b*, Gall-nuts of Cynips quercus folii.

ployed to produce ink, dyes, gallic acid, etc.; but Berzelius assures us that they contain little more tannic acid than the leaf itself on which they are produced.

Messrs. Lacaze and Riche ("Archives des Sciences Physiques et Naturelles de Génève," xxx. 17) have profited by the singular conditions under which the young *Cynips* are developed in the gall-nuts, to solve an important physiological problem:

As grease exists in the vegetable as well as in the animal world, it was an interesting question to know whether animals derived their fat wholly from vegetables, or whether this substance could be formed in the animal body. The vegetable tumours in which the larvæ of the *Cynips* are found contain no grease or oily matter, whilst the grub that grows in them is remarkably fat! It is evident, therefore, that animals have the power of forming fat or grease by means of the starch or other principles supplied by vegetables.* The conditions under which fat is most readily formed are indeed those in which the larvæ of the cynips live, namely, a vegetable or farinaceous diet, repose, solitude, and obscurity.

It is not improbable that other insects besides kermes, coccus, and cynips may become important as dye-producers. Reaumur has spoken of an aphis which produces galls in different parts of Asia, and these galls are employed to dye silk a crimson colour. Linnæus also speaks of the tinctorial properties of *Aphis pini*, an insect common in our climate, and which produces a sort of gall-nut at the extremities of the spruce fir. When these galls have attained their maturity, says he, they burst and discharge a

---

* Dumas and Milne Edwards formerly arrived at the same conclusion. They fed bees exclusively upon honey and sugar, and found that they produced *wax*, an observation which Huber had already made many years before.

yellow powder, which stains the clothes. A tree common enough in India, and which is called *Terminalia citrina*, yields a sort of gall, which serves in that country as a dye; to it indeed the natives owe their best and most durable yellow colour. It is produced by a hitherto unknown insect. Among the little "money-spiders" (*Trombidium*) which attract the attention of children in the garden about spring, *Trombidium tinctorium* is used in Guinea and Surinam as a dye. I have observed that when acid vegetable colours of a yellow tint can be fixed upon silk, cotton, wool, etc., they can almost always be turned crimson by alkalies. It is impossible yet to say what influence the newly discovered colour *magenta* will have upon the cochineal production. But as *carminium* and *magenta* are so very different in properties, it is probable that the production of magenta dye will not materially affect that of cochineal.

Chapter IV.

Insects producing Wax, Resin, Honey, Manna.

Chinese Coccus which produces a kind of Spermaceti—Value of its Produce—White Lac—Insects producing Resin—Grey-wax Insect of Sumatra—Details concerning the wax Coccus—Bees—Apis mellifica—Its native country—Virgil—Modern Authors who have written on Bees—Apis ligustica—A. amalthea and its curious Nests—Bamburos—Apis unicolor—Green Honey of Bourbon—Rock-honey of North America—Apis fasciata—A. indica—A. Adansonii—A swarm of Bees—The Queen, Males, and Workers—Mathematics of the Bee-cell—Silk produced by Bees—Production of Wax—How Honey is procured—Plants favourable to Bees—Duration of life in Bees—Enemies and Maladies—Chloroforming Bees—Mr. Nutt's Hives—Profit derived from Bee-culture—New modes of Preserving Bees during winter—Periodical transportation of Hives—How to discover Bees' Nests—New species of Bee at Sydney—Bees as Instruments of War—Honey, its Nature and Composition—Artificial Honey from Wood, Starch, etc.—Manna and the Coccus Maniparus—Wax—Its Nature, Composition, and Uses.

INSECTS PRODUCING WAX, RESIN, HONEY, MANNA.

WE must again turn to the genus *Coccus*, to speak of a species of wax-producing insect which is attracting particular attention in France at this moment. This will be better understood when it is known that the French pay four millions of francs annually for wax; and the *Coccus* of which I speak produces about ten millions of francs' worth of wax per annum. It is a Chinese insect, and the wax it produces resembles spermaceti. It was first alluded to by Grosier, who remarked that towards the beginning of winter small tumours appear on the trees it inhabits. These tumours increase in size until they are as large as a walnut. He imagines these to be the nests of the female insects; they are filled with eggs which hatch in the spring, and the young insects disperse themselves on the leaves and pierce the bark. The wax they produce—probably in the same manner that *lac* is produced by *Coccus lacca*—is perfectly white, and known to the Chinese as *Pe-la* (white wax). It begins to appear about June, and is gathered by the natives at the beginning of September. The quantity produced in China alone is, according to Geomelli Careri, sufficient to supply the whole nation with this useful article. This insect, with whose specific name we are not yet acquainted, is cultivated chiefly in the province of Xantung, like the cochineal in that of Oaxaca, and there its breed has attained great perfection; but it is also reared with more or less success from the frontiers of Thibet to the Pacific Ocean. The plant on which it lives is a species of privet, *Ligustrum lucidum*, a Chinese shrub.

The chemical examination to which this wax has been submitted, proves it to be superior to any yet discovered, and shows that it bears a close resemblance to spermaceti.*

From what precedes it will be seen that the acclimatization of this insect in France becomes an exceedingly interesting problem. It appears probable, from observations we already possess, that the Chinese spermaceti *Coccus* is not confined to China, and that it, or at least some analogous insects producing wax, are found in other parts of Asia. Dr. Anderson formerly described as *white lac* a substance similar to the white wax of the Chinese *Coccus*, and

---

* This Chinese wax must not be confounded with that called *vegetable wax*, produced by palms and by several species of *Myrica*, etc. (On these see Cook in the "Technologist," London, June, 1861.)

which, he said, could be produced in any quantity, near Madras, at a much cheaper rate than beeswax. And from De Azara's observations, a similar wax-producing *Coccus* appears to abound on a small shrub in South America.

So many trees (*Palms*, and *Myrica*, and *Rhus* especially) are known to produce excellent wax without the aid of any insect, that we cannot always decide at first whether this substance is the product of the plant or of the insect.

Molina has shown that at Coquimbo in Chili large quantities of resin are produced by several species of the shrub *Origanum*, as a consequence of the bite of an insect. The latter is a small red caterpillar which changes into a yellowish moth with black stripes on its wings (*Phalœna ceraria*, Mol.) Early in the spring vast numbers of these caterpillars collect upon the branches and buds of the tree, where they form cells of a kind of white wax or resin; and in these cells they undergo their metamorphoses. The wax, which at first is very white, becomes gradually yellow and then brown. It is collected by the inhabitants in autumn; they boil it in water, and make it up into cakes, which go into the markets. They use this wax instead of tar for their boats.

There exists at Sumatra a species of winged ant that produces a sort of grey wax. A sample of this substance was exhibited at the French Exhibition of 1855, but we have as yet no details concerning the insect that produces it.

All the insects of the genus *Coccus* contain a considerable amount of grease, from which *stearine*, the element of our modern " wax-candles," has been extracted ; moreover, Berzélius extracted from *Coccus polonicus* the acids which are contained in butter ; and it is probable that butyric acid exists in the whole genus.

The latest information we have concerning the spermaceti *Coccus* of the Chinese we owe to M. Stanislas Jullien, who ascertained in 1840 that these insects were cultivated indefatigably by the Chinese, on three different sorts of plants, with equal success; namely, the plant they call *nint-ching*, which M. Brogniart tells us is the *Rhus succedanea;* the *tong-tsing*, which Thunberg says is *Ligustrum glabrum;* and the *goukin*, a plant which grows in damp places, and is probably the *Hibiscus Syriacus*, or belonging to the same family as the latter. The wax which is obtained from these trees abounds in all the east and south provinces of China. It is collected by scraping the trees in autumn, it is then boiled in water, and strained through a cloth, after which it is placed in cold water, when it becomes solid, and then resembles soap-stone or steatite. The young insects, according to M. Stanislas Jullien, are hatched from eggs of a considerable size, and cover the trees about June. They are soon observed to secrete a sort of viscous liquid, which adheres to the branches, and transforms itself slowly into a kind of grease or white wax. In September this grease adheres so firmly to the branches that it is difficult to remove it. The more sap the tree yields the more wax the insect produces; it would, therefore, be interesting to try the effects of some of our artificial manures upon these trees and their insect burden. The insect appears to nourish itself upon the sugar contained in the sap, which it transforms into a liquid grease, becoming solid on contact with the air. Although insects are certainly instrumental in causing the production of several varieties of wax, it is not proved that they promote the formation of the Japan wax furnished by *Rhus succedanea*, a plant extensively cultivated in Japan and China. The wax of this shrub is now being imported in England in enormous quantities.

I must now allude to bees. I really dread the task of saying anything about these insects, so familiar to all, and upon which so many useful and instructive volumes have already been written; but on account of their utility to man, bees have long since been placed upon the first rank among *domesticated* animals. An ancient historian, Niebuhr, states that he met between Cairo and Damietta a convoy of 4000 hives, which were being transported from a region where the season for flowers had passed, to one where the summer was later.

Our domestic hive-bee (*Apis mellifica*, Fig. 6) appears to be a native of Greece;* from whence it was subsequently introduced into the different countries of Europe. It is a well-known fact that the education or rearing of bees attained to great perfection among the ancient Greeks, more especially among the inhabitants of Attica; the honey of the latter country was always considered extremely fine. Ancient philosophers looked upon bees as forming part of the universal soul of the world, and believed that the sweets upon which they lived made them participate in divine nature; thus, we see the ancient poets celebrating the works of the bee, making known their habits and writing their history. It was from these sources that Virgil collected ideas, added to them the results of his own observations, and produced the charming verses of his "Georgica."

Fig. 6.—*Apis mellifica* (Hive-bee).

Among the moderns the following are the names of distinguished entomologists who have written considerably on bees:— H. Huber, P. Huber,

* Most authors agree upon this point.

Reaumur, Bonnet, Latreille, Needham, Kirby, Swammerdam, Kirby and Spence, Mills, Thorley, Hunter, Keys, Bonner, Schiroch, Bevan, etc., etc.

Apis mellifica, the domestic bee, reared in **hives**, is the same throughout Europe, except in some parts of Italy, the Morea, and some of the Grecian isles, where another species is cultivated, the *Apis ligustica* (?) of Spinola. The domestic bee (*A. mellifica*) is found wild in the forests of Russia, and some parts of Asia, where it builds its nests in hollow trees. Another kind of bee, the *Apis amalthea* of Latreille, is found at Cayenne, where it builds curiously-shaped nests upon the tops of high trees; these nests are something like a bagpipe. They are seen also in South America, and furnish large quantities of honey, but this honey, though very sweet and agreeable, is very liquid and difficult to keep, as it easily ferments.

Another species of wild bee, which has been called *Bamburos*, is very plentiful in the woods of Ceylon, where it is eaten as a delicacy, though it furnishes a considerable harvest of honey to the peasants.

In the Ukraine some of the country people, we are told, derive more profit from the sale of their honey than from their corn; some peasants keeping as many as 500 hives each. The Indians of Paraguay, the natives of the Isle of Bourbon, of Madagascar, etc., live, to a great extent, upon the honey of the bee. The honey exported from the Isle of Bourbon is the product of *Apis unicolor,* Latreille ; it is of a green colour and oily consistency, and has an aromatic flavour.

In North America there is a bee which suspends clusters of thirty or forty wax cells, resembling a bunch of grapes, to the rocks. Its honey is called *rock-honey.* It is very clear and thin, somewhat like water.

The honey contained in the hives that Niebuhr met upon the Nile was the product of *Apis fasciata,* a species of bee extensively cultivated in Egypt.

Apis unicolor has been domesticated in Madagascar; *Apis indica* is educated in some parts of India; and *Apis Adansonii* has been extensively reared in Senegal.

Although in Spain the number of hives is very great—we read of an old parish priest who had 5000 ! —in France the cultivation of the bee is not so much attended to.

The honey of *Apis mellifica*, L., is imported (from Europe, Asia, and America, chiefly from Lisbon) to Liverpool, at the rate of about twenty-seven tons a year. Wax is imported from Europe, Asia, Africa, and America, at the rate of twenty-five tons per annum into Liverpool alone.

Until very recently,* nearly the whole of the wax employed in Europe, and most of that consumed in America, was the produce of the hive bee.

A swarm of bees is composed of one female (generally known as the queen-bee), from 600 to 1200 males, and from 15,000 to 30,000 working bees, which have no sex. Aristotle used to call the chief of the hive the *king*-bee. The working-bee would have become a female had it attained its perfect development—a fact discovered by Mdlle. Jurine, a lady who first dissected the working-bee; but whilst in the larvæ state, being fed upon a small allowance of food, and bred in small cells, its growth is impeded, its ovaries avort, and it comes forth definitely as a working-bee.

The female (the queen) only comes out of the hive or nest upon two occasions: the first at the period of coupling, when she soars in the air with a host of males, one of which is finally chosen as her mate. This one dies almost immediately afterwards, and the female returns to the hive. The queen-bee has thus become fertile for one year—often for her whole life. As soon as the males return to the hive they are unmercifully put to death by the working-bees. The male-bees (drones) have no sting. This takes place about August.

* At present there is a considerable importation of *vegetable wax*.

Forty-eight hours after the female bee has returned to the hive she begins to deposit her eggs in the cells destined to receive them. During the first summer few eggs are laid (principally those from which "workers" emerge). In winter the laying ceases, to re-commence in the spring, when, in about three weeks, more than 12,000 eggs are deposited by the same queen-bee, which begin to hatch in three or four days.

In a single season a queen-bee will sometimes lay from 70,000 to 100,000 eggs. Reaumur says that upon an average she will lay 200 in a day.

The queen-bee must be eleven months old before she can produce eggs which produce males, and still older before the eggs she lays will bring forth female bees.

The second occasion on which the female-bee leaves the hive or nest is when a new female has been born, and emigration becomes necessary. It is then that *swarming* takes place. When a swarm issues from the hive, it is customary among the peasants to make a noise, to throw sand into the air, and to imitate a storm. The bees then fix themselves in a cluster to some object, from which they are shaken into the new hive.

One word upon the queen-bee. She is always born in one of the *royal cells*, which are larger than the others. She receives a particular kind of nourishment while in the larva state, and if by any accident the queen-bee of a hive is lost or killed, the remaining bees have the power of nourishing any of their common larvæ in such a manner as to produce a queen.*

A word upon the working bees. There are two varieties: the *wax makers* and the *nurses*. The former are large and robust, they fly into the country to collect the pollen and sugar of flowers; the others, less strong, remain in the hive; their duty is to feed the young larvæ.

A beautiful example of applied mathematics is furnished by the bee-cell. Each cell of the honey-comb is a hexagon, the base of which is composed of three rhomboidal plates so composed as to contain the largest amount of honey with the least quantity of wax.†

Lord Brougham, in a paper read at the Paris Academy (May, 1858), asserts that the cells of the larvæ of bees are lined with a species of silk; when the wax is separated there remains behind what appears to be a very fine tissue of silk.

It is now beyond doubt that the wax of the bee is not taken from the vegetable world, but is produced by the insect itself. The fact was ascertained

---

* See on this Kirby and Spence "Introduction to Entomology." Lond, 1858, pp. 361, 362, *et seq*.

† See Kirby and Spence, *loc cit*, p. 273.

by Thorley in 1744, and afterwards by Huber, who described the organs, situated on each side of the abdomen, which secrete the wax in the shape of thin plates.

Honey, on the contrary, consists of the sugar which is taken directly from the nectaries of the flowers. It is lapped up from these curious parts of the flower by the tongue of the bee, and transmitted into the first stomach or honey-bag of the insect. It is never found in any other part of the bee's body. When the insect is laden it returns to the hive, and disgorges the honey into cells which are destined to receive it.

Plants which are peculiarly adapted to the bee are species of *Echium, Borago, Verbascum, Thymus*, and the *Crucifera*. In some countries bees attach themselves to particular plants; for instance, in the Highlands of Scotland and in Sweden, to the *Erica*, or heath-plant; in Scania, to the buckwheat; in Poland, to the lime-tree; in Narbonne, to rosemary; in Greece, to thyme; in Corsica, to the arbutus; in Sardinia, to the Artemisia, etc.; and hence arises the different flavours and qualities of honey in the several European markets. Other plants appear to be avoided by bees: thus the poisonous nectar of the oleander, which proves fatal to thousands of flies, will not be touched by the bee. But a few cases are on record of bees gathering poisonous honey, and causing extensive mortality among those who eat it.

The duration of the life of bees has been a subject of controversy. Virgil and Pliny say seven years, other writers ten; but of the five hundred bees which Reaumur marked with red paint in the month of April, not one was living in November; and more modern authors state that the working bees are *annual* insects, but that the queen may live two years. We have already seen that the males die every year. However, by a succession of generations hives have been preserved for more than five and twenty years; and Thorley states that a swarm of bees that took possession of a spot under the leads of the study of Ludovicus Vives, in Oxford, in 1520, were still there in 1630. They had therefore propagated their race in this spot for a period of one hundred and ten years.

The enemies of bees are mice, rats, swallows, and other insectivorous birds, wasps, ants, and some other insects. They are also subject to certain diseases, such as dysentery, indigestion, etc. Hives should be placed in a quiet spot, away from noise; if wasps' nests exist in the neighbourhood, they should be destroyed; ants' nests likewise; and frogs, toads, ants, spiders, etc., must be kept away. Bears and foxes are very fond of honey. When a person approaches a hive, he should speak *mezza-voce*, as the

Italians say; and if the bees appear hostile, he will do well to stoop down. Liquid ammonia is employed with success to cure the effects of their sting.

Mr. Nutt's system of hive appears to be held in esteem upon the Continent. It is no longer necessary to kill these useful insects in order to procure their honey, as every apiarist knows they may be fumigated or "chloroformed" in different ways. The fumes produced by burning *fungi* permit the cultivator to attain this end without the loss of his bees. Of these fungi the common puff-ball (*Lycoperdon*) is to be preferred; its fumes act upon animals like chloroform, as Dr. Richardson has proved by several experiments. The asphyxiation of bees by the puff-ball fungus has been practised by Messrs. Blondel and Cossart with success, thus: A hole is made in the earth a few inches deep, and wide enough to hold a plate, under which is placed a towel. Four or five puff-balls, perfectly dry, are passed on to a long iron pin and lighted. The pin is then stuck into one of the sides of the excavation, and the hole covered with the bee-hive, the ends of the towel being pulled up and fastened against the hive by the loose earth, the smoke is prevented from escaping. In four or five minutes the hive may be lifted up; all the bees are found upon the plate in a state of insensibility. This operation is best performed at about four o'clock in the afternoon. When the bees are again placed in the hive, the opening of the latter is nearly closed, so that they may not make their escape when animation returns. The next morning they are permitted to go out, and are as lively as before. But Mr. Nutt's system of hive, where the honey is taken from the top, without suffocating the bees, renders this operation unnecessary.

The profit derived from the cultivation of bees has been often much exaggerated. Large fortunes are not more easily realized by this undertaking than by other means. Bees require a great deal of attention, and to realize a profit at all the cultivator must, in most cases, submit to a considerable amount of trouble, and often to no little anxiety.

The sales of *swarms*, *wax*, and *honey* are the three elements or basis upon which bee-culture rests. The best time for purchasing swarms is in the month of October. On honey and wax we shall say a few words presently.

The production of a hive depends principally upon the mildness of the climate. In the environs of Paris there are bee-hives which realize a pure profit of twelve to twenty-four francs a year. These figures may be taken as a sort of *criterion* in our climate. Those who occupy themselves with the rearing of bees should possess " Les Observations sur les Abeilles," by H. Huber, of Geneva; "Les Nouvelles Observations," by the same author, noted by P. Huber; also the works of Reaumur, and those of the English authors whose names we have already mentioned.

The principal losses experienced in bee-culture occur during the winter; they arise either from the bee-keeper having, with a miserly hand, deprived the insects of too much honey, or from a bad mode of preserving the hives through the winter season.

1st. To ascertain whether a sufficient supply of honey has been reserved the average weight of the hives must be consulted.

2nd. M. Pénard-Masson, a French apiarist, assures us that he has derived considerable benefit and preserved throughout the winter hives which otherwise would have perished, by turning a certain number of bees out of a hive where the supply of honey is too small, into one where there exists an excess of nourishment.

But one of the newest and most original methods of preserving bees during winter is that lately discovered by M. Antoine of Rheims. His process consists in *burying* the hives with great care, and as quietly as possible. About the 15th of November, a ditch, a good depth, and wide enough to contain all the hives that are to be interred, is dug in the middle of a field, away from any road or thoroughfare.

The hives are placed in it with the utmost care, avoiding as much as possible motion and noise. Their sides are protected with boards and straw, and the whole is then covered with the earth removed in digging the ditch. Seeds are immediately sown over the spot, to hide more completely the buried treasure. The excavation is opened on the 15th of February following, and the bees removed with the same care as before. These operations are executed in the evening.

By this system, it appears that the bees consume three-fifths less nourishment than if they had not been buried, the mortality in the hives is almost *nil*, and the queen begins to lay three weeks sooner than usual. I should imagine that porous ground should be chosen in preference to a heavy clay soil, for burying the hives.

Mr. Newport in his paper published in the "Philosophical Transactions" for 1837, has proved that in our climate bees are never, strictly speaking, torpid during the winter season, but preserve throughout it a certain degree of activity.

Towards the end of October, when the inundations of the Nile have ceased, and the peasants can sow their land, sainfoin (*Hedysarum*) is one of the first plants sown, and as Upper Egypt is warmer than Lower Egypt sainfoin flowers first in the former district. At this time, according to Kirby, bee-hives are transported in boats from all parts of Egypt into the upper district, and are then heaped in pyramids upon other boats prepared to receive them. In this station they remain some days, and are then removed lower down, where they remain the same time; and so they proceed until the month of February, when, having traversed Egypt, and arrived at the sea, they are dispersed to their several owners. A similar transportation of hives occurs in Persia, Asia-Minor, Greece, sometimes in Italy, and even in England in the neighbourhoods of heaths.

The honey-hunters of New England seek the wild bees' nests in the following manner:—Whilst the sun shines brightly a plate containing honey is set upon the ground. It soon attracts the bees, who feed greedily upon it until their honey-bag is filled. Having secured two or three that are thus satiated the hunter allows one to escape. The insect rises in the air, and being completely laden, flies straight towards its nest. The bee-hunter then strikes off for a few hundred yards at right angles to the course taken by the first bee, and lets fly another; he observes its course with his pocket compass. The point where the two courses intersect each other is the spot where the nest is situated.

The *bulletin* of the Paris "Société d'Acclimatization" for 1856 announces the discovery of a new species of bee (*Apis*) at Sydney. It inhabits the hollow portions of decayed trees, lives together in prodigious numbers, appears to have no sting, and produces a brown-coloured wax, and an excellent description of honey. This is all we know of it at present. If it has no sting, it is probably not an *Apis*.

In time of war, the ancient Egyptians used to place implicit trust in their *sacred beetles*; but bees have been employed as more efficacious instruments of war. Lesser reports that in 1525 a mob of peasants, who endeavoured to pillage the house of a gentleman, were dispersed by the servants of the latter, who flung some ten or twenty bee-hives into the mob. We have read somewhere than an American slave ship was boarded and captured by means of bee-hives.

Honey is formed from the sugar secreted in the nectaries of flowers. It is composed of two distinct kinds of sugar, known to chemists as grape-sugar and liquid sugar, which both differ essentially from cane or beet-root sugar, though their composition is similar. They are less sweet than the latter. Liquid-sugar cannot be made to crystallize like the other varieties.

The sweet liquid extracted from the nectaries of flowers possesses most of the properties we observe in the honey of the bee. Some flowers contain a considerable quantity, such are, for instance, the trumpet-honeysuckle, whose sugar is out of the bee's reach, and the *Cobæa scandens*, each flower of which contains almost enough sugar to sweeten a cup of coffee.

But there is an important difference between honey and the sweet juice of the nectaries of flowers. The former contains no cane-sugar, whilst the latter, as Braconnot has shown, yields by evaporation some crystals of cane-sugar. The *Rhododendron ponticum* and the *Cactus Akermanni* were found to contain so notable a proportion that one corolla of the latter gave as much as one-tenth of a gramme of crystallized cane sugar. It is evident, therefore, that this cane-sugar of flowers is converted into grape sugar in the honey-bag or the cells of the bee.

When honey is allowed to stand for some time, it gradually thickens and consolidates. By pressure in a linen bag it may then be separated into a white solid sugar—called grape sugar, as it is found in grapes and raisins—and a thick semi-fluid syrup, called liquid sugar. Grape sugar is better extracted by placing the honey upon a porous brick, which absorbs all the liquid sugar, whilst the grape sugar crystallizes at the surface.

The liquid sugar of honey often contains odoriferous substances produced by the flowers from which it has been extracted. To these the honey owes a certain fragrance or flavour for which it is much prized. Such is the case with the honey of Mount Ida, in Crete; hence also the perfume of Narbonne honey, of the honey of Chamounix, and of our own moorland honey when the heather is in bloom.

Honey is extracted from the comb by gently heating the latter and letting as much as possible run out. When no more can be extracted in this manner, the comb is again gently heated and pressed. Hence two distinct qualities of honey. The comb which has been pressed is treated with water, and furnishes a liquid which, on being fermented, produces *hydromel*, a sort of vinous liquid employed in medicine. Finally the combs are placed in sacks and submitted to the action of boiling water to obtain the wax. Honey is employed as an agreeable aliment; it is used in various forms for medicinal purposes, and enters into the composition of gingerbread.

Honey can be artificially made by boiling wood, linnen, cotton, or starch in water acidulated with sulphuric acid. The liquid is allowed to boil from ten to twenty hours, and the water replaced as it evaporates. The acid liquid is then saturated with chalk, filtered, and evaporated, when a syrup resembling honey is obtained. This syrup is indeed composed of grape sugar, mixed with a small quantity of liquid sugar; and this, as we have seen, is the composition of honey. This discovery is owed to Braconnot.

Mannite, the sweet principle of manna, has been found, though rarely, in some kinds of honey.

The manna that is used as an agreeable food in the East, and with us as a purgative for children, is caused to flow from the *Tamarix mannifera* (Fig. 7), by the punctures of a small insect, *Coccus maniparus*. But it is essentially a vegetable product,

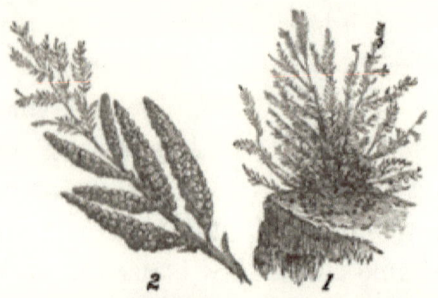

FIG. 7.—Tamarix mannifera (Manna-bearing Tamarix).
1. Shrub twelve feet high. 2. Branch with fruit.

being obtained from the sap of the ash tree (*Fraxinus ornus*, *F. rotundifolia*, etc.). The little green *aphides* of the lime tree appear, however, to secrete mannite from their bodies, on account of which they are captured and reared by ants as we breed cows for their milk. But it has not yet been proved that any animals produce mannite directly, though sugar is a common product of the animal economy. Besides the different varieties of ash, the tamarix, and seaweeds,* a sort of manna is produced in Australia and Van Diemen's Land by the *Eucalyptus resinifera*. At certain seasons of the year a sweet substance exudes from the leaves of this tree, and dries in the sun, and when the wind blows hard enough to shake the trees, the manna falls like a shower of snow. Certain oaks, larches, pines, cedars, etc., produce a similar substance. The cedar-manna, which is brought from Mount Lebanon, is the product of *Pinus cedrus*—it sells for twenty or thirty shillings an ounce. The manna collected by the Arabs for food in the desert, is the product of *Hedysarum alhagi*, L., a plant which is indigenous over a large portion of the East. That of Mount Sinai is obtained from the *Tamarix* before alluded to. The *Coccus manniparus* infests this tree, from which the manna exudes as a thick syrup, which, during the heat of the day, falls in drops, but during the night congeals and is gathered in the cool of the morning.

On beeswax I have little to say. The best and whitest wax is that taken during the month of March. The nature of wax has been very completely investigated by Dr. Levy of Paris, to whose admirable paper ("Annales de Chimie," xiii. p. 438)

* On the production of Mannite by seaweeds, see my paper in "Comptes Rendus," Paris, 1st Dec., 1856.

I must refer my readers. We have already seen how it is produced by the bee, the Chinese *Coccus*, and the manner in which it is extracted from the honeycomb. We have also seen that wax is produced by many vegetables, amongst others by the cabbage; it is also found in the *pollen* of flowers, from which it was long supposed the bees procured it. But the wax contained in *pollen* differs from beeswax; it is the substance known as *propolis*, which the bees use to fill up fissures in the nest or hive. The wax of the honeycomb can be separated into two distinct substances by means of spirits of wine; the first, called *cerine*, dissolves in boiling spirit, and the liquid on cooling deposits it in white gelatinous crystals. The substance which remains undissolved is *myricine*, which does not crystallize.

Wax is still employed in considerable quantities (in spite of the discovery of stearine candles) for candles used in Roman Catholic churches. It has of late years been notably employed in photography, to wax the paper and render it translucide. The wax produced by certain wild bees, called *Mellipona*, and gathered at Costa Rica, in the Island of Cuba, etc., has lately been applied to the manufacture of lithographic ink. Finally wax is employed for an infinite number of minor uses, for making anatomical models, busts, dolls, etc.

Chapter V.

Insects Employed in Medicine, or as Food, and other Insects useful to Man.

Spanish Flies—Cantharides—Their Medical Properties—Cantharidine—Cantharides in Poitou—Different **Species of** *Cantharides—Discovery of Cantharidine in Meloe—The Meloe, or Oil Beetle—Metamorphoses of Meloe and Sitaris—Cetonia Aurata—Coccinella—Trehala—Buprestis—Ants—Formic and Malic Acids in Ants—Production of Milk from the Eggs of Ants—Ants which collect Precious Stones—Termes as an Article of Food, etc.—Locusts and Cicadæ—Acrydium migratorium—*The **Ethiopian** *Acrydophaghi—Cicada septemdecim—***Bugs** *and Fleas—Southey—Phtirophaghi—Aranea edulis—Centipedes—The Mexican Boat Flies—Beetle used for Soap—Calandra granaria—Presence of Tannic and Gallic Acids in this Beetle—Fire Flies—Truffle Flies—The Common House Fly, etc.—Remarkable Action of Light upon Animal Life—Growth of* **Insects** *under the Influence of differently Coloured* **Light.**

INSECTS EMPLOYED IN MEDICINE, OR AS FOOD, AND OTHER INSECTS USEFUL TO MAN.

ONE of the most important insects, in a medical point of view, is the beetle called Spanish fly (*Cantharides*), of which there are many species, all dangerous poisons. They are employed outwardly for their blistering and exciting properties, and inwardly, for various disorders, as an energetic stimulant.

Their poisonous action manifests itself by violent irritation of the membranes of the stomach and intestines. The vesicatory or blistering property of these beetles is owing to *Cantharidine*, a principle extracted from them by Robiquet, and studied by Gmelin. They contain also a peculiar volatile oil, mentioned by Orfila, but of which little is yet known, except that it appears to be this oil which gives *Cantharides* their peculiar odour.

Cantharidine crystallizes in small white crystals, soluble in ether and boiling alcohol. This substance is only capable of producing inflammation or blistering; the exciting or aphrodisiac action of *Can-*

tharides is owed to some other principle as yet unknown, as Schroff has lately shown.

M. Babinet has informed me that in some parts of France, more especially in Poitou, ash-trees are never planted, because the quantity of *Cantharides* that breed upon these trees soon becomes intolerable to the inhabitants of the district.

In our climate, *Cantharides* are to be found upon the lilac, the privet, and some other shrubs. They are very plentiful in Spain (hence their appellation, " Spanish fly"), Italy, Sicily, etc., but comparatively rare in England, where they are only to be met with now and then in the southern counties.

Of these beetles, the *Cantharides vesicatoria* of Geoffroy and Latreille is most frequently found in commerce; it is distinguished by its strong and peculiar odour, its wing-sheaths or elytra of metallic green, and its black antennæ or horns. In America, two other species, namely, *Cantharides cinerea* and *C. vittata*, being extremely common and noxious insects, are more frequently used than *C. vesicatoria*. In India, *C. gigas* and *C. violacea* are employed; in Sumatra and Java, *C. rificeps*; in Brazil, *C. atomaria*; in Arabia, *C. syriaca*; in China, certain species of *Mylabris*, a genus closely allied to *Cantharides*.

The real Spanish fly, *C. vesicatoria*, Latr., is imported into Liverpool from Italy at the average rate of three hundredweight per annum.

Our readers are probably all acquainted with the *Meloë proscarabœus* (Fig. 8), or oil-beetle. It derives its name from the fact that, when taken into the hand or otherwise irritated, it secretes a fragrant oily fluid, to which have been attributed the most wonderful qualities; amongst others, that of infallibly curing rheumatism! This large beetle is easily recognized by its dark violet colour, its elytra, which are oval, and so short that they do not cover more than one-third of the insect's body. Late in the spring, *Meloë proscarabœus* is often seen in our fields and on the hedgebanks, drawing its heavy body slowly over the damp grass. To preserve it in insect collections, its body must be stuffed with cotton-wool, otherwise it shrinks to an incredibly small bulk.

FIG. 8.—Meloë proscarabeœus (Oil-beetle).

Sobrero and Lavini have recently discovered *Cantharadine* in insects belonging to this genus *Meloë*, which is closely allied to the genus *Cantharides*, more so, indeed, than that of *Mylabris*, mentioned above. In Spain, these oil-beetles, or *Meloë*, are still used in lieu of Spanish fly.

M. Fabre, a very distinguished entomologist, has recently made known some facts relating to *Meloë* and the allied genera *Sitaris*, which are so curious that I think they may safely be related here :—

The insects belonging to the two genera, *Meloë* and *Sitaris*, together perhaps with the whole tribe, are, in their early stages of life, parasitical insects, living upon the bodies of certain honey-making *Hymenoptera*. From M. Fabre's account, it appears that their *larvæ, before arriving at the pupa or chrysalis state*, go through no less than four distinct metamorphoses. The author finds himself obliged to invent new names to designate these newly-discovered phases of insect life. He therefore denotes them *primitive larva, second larva, pseudo-chrysalis*, and *third larva*. The passage of one of these forms to the other is effected by a simple process of moulting, or throwing off of the outer skin; the viscera remaining unchanged.

The primitive larva is a hard, crusty little being. It lives on the bodies of Hymenoptera (bees, etc.) until it is transported to the nest and finds itself deposited in the bee-cell. Once there it soon devours the offspring of the Hymenoptera. The second larva, which is developed in the cell, lives upon the honey. It is much softer than the former. The pseudo-chrysalis resembles a piece of hard gutta-percha, it is quite devoid of motion, its sheath is of a hard horny substance, upon which can be observed the rudiments of a head and six small tubercles, rudiments of feet. The *third larva* bears a strict resemblance to the *second larva*. From this stage the usual metamorphoses of insect life begin, and follow out their ordinary course: this third larva becomes first a chrysalis, from which it emerges as a perfect insect.

Other coleopterous insects are endowed with inflammatory or blistering properties. Such, for instance, is the *Cetonia aurata,* or golden beetle, which was employed in the time of Pliny, and which plays such an ingenious part in the tale of Edgar Poe. Such again are the *Coccinella,* or lady-birds, which, when captured, secrete from their legs an acrid yellow fluid having a disagreeable odour. It is doubtless to this fluid that they owe their property of curing the most violent toothache when they are placed alive in the hollow part of the tooth.

A pharmaceutical substance, known as *Trehala,* has lately been studied by M. Guibourt. It is a kind of insect-nest or hollow cocoon, round or oval, about the size of a large olive, and is the produce of a coleopterous insect (or beetle) closely allied to the genus *Curculio,* and named *Larinas nidificans.* This insect lives on the branches of a shrub, a species of *Echinops.*

The trehala is composed of 66·54 parts of starch, 4·66 of a kind of gum, and 28·80 of sugar, mixed with a small quantity of some bitter principle, and mineral salts. In the East this substance is as much used as salep or tapioca. It was first noticed in Syria. When placed in water it swells considerably, becomes soft, and finally transforms the liquid into a sweet mucilaginous decoction. M. Berthelot has just extracted a new kind of sugar from this cocoon. It resembles cane-sugar to a certain extent, and has been called *trehalose*.

The wing-cases or elytra of that beautiful Indian beetle, *Buprestis vittata*, are occasionally imported from Calcutta to Liverpool. They have a bright metallic green lustre, and are employed to ornament *Khus-khus* baskets, fans, etc., and on muslins to enrich the embroidery. *Khus-khus* or *vitiver* is the dried root or *rhizome* of a grass, *Andropogon muricatus* (Retzius). This sweet-scented root arrives here now and then from India. It is made into baskets, fans, mats, *sachets* for the wardrobe, etc.; which are often most sumptuously decorated with the wings of *Buprestis vittata*.

Ants (*Formica*) are useful insects in a variety of ways. By distilling them a peculiar substance called *formic acid* passes over—but it can be obtained with greater ease and economy from starch* — in the residue that remains is found a certain proportion

---

* By distilling starch with dilute sulphuric acid and peroxide of manganese.

of *malic acid*, an acid first discovered in the apple.

Certain large ants, called *Cupia*, in the Brazils are eaten by the natives, and so is another large species called *Tamajoura*. In Africa ants are sometimes stewed with butter, and considered delicious. In Sweden they have been distilled with rye to give a peculiar flavour to brandy. By submitting ant-eggs to pressure, the chemist John produced a kind of milk resembling a mixture of milk and chocolate. This liquid, upon analysis, was found indeed to contain albumen, lactic acid, phosphoric acid, a matter resembling casein, and a yellow grease like butter, so that its composition as well as its taste resembles that of ordinary milk.

Ants are also very useful to **medical students,** in preparing skeletons of small animals, such as moles, rats, etc. The dead body of any of these animals being placed in or near an ants' nest is soon reduced to a very clean skeleton. Other insects might also perhaps be used for this purpose.

On the high plateaux of the Rocky Mountains, according to Humboldt, there exists a species of ant, which, instead of useing fragments of wood and vegetable remains for the purpose of building its dwelling, employs only small stones of the size of a grain of maize. The instinct of the insect leads it to select the most brilliant stones for this purpose, and these ant-hills are frequently filled with transparent quartz and garnets. At Capula Humboldt found the ant-hills filled with shining grains of obsidian and sanidine.

Ants belong to the family of *Hymenoptera* (bees, cynips, etc.); but there are insects called *white ants* (*Termes*) which belong to the family of *Neuroptera* (dragon flies, ephemera, etc.). The latter are very useful to man in certain parts of the world as an article of food, though they certainly are most terrible enemies to our habitations and furniture. In France there are numerous examples of old houses, or large pieces of furniture falling in, as a consequence of the mining operations of the *Termes*. De Quatrefagés recommends us to destroy them by means of a current of chlorine gas directed into their galleries, as Thenard once effected the destruction of the rats of Paris by means of sulphuretted hydrogen.*

In the torrid zone, where the *Termes* abound, they build nests like hills, eleven or twelve feet high, which are often mistaken at a distance for the huts of the natives. Their habits are as interesting

---

* The British Government has lately applied to the Entomological Society to know the best means of destroying the white ants which infest certain of our colonies. Several remedies (arsenic-soap, lime, corrosive sublimate) were hinted at by the members, but chlorine was not mentioned.

as those of bees; but we must refer our readers to special works on entomology for a description of these. The Hottentots eat them boiled or raw; they serve as food in the East Indies. The Africans roast them in iron pots and eat them by handfuls, as we do sugar-plums. They resemble in taste (according to Smeathman) sugared cream, or sweet almond paste. They constitute an extremely nutritious article of diet.

Many parts of the world, and great portions of Europe are often ravaged by certain species of locusts, chiefly by the species *Acrydium migratorium* (Fig. 9), which I have found as far north as Ostend

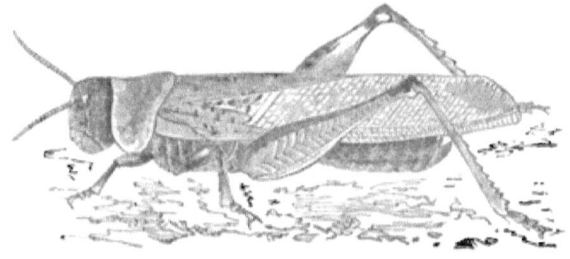

FIG. 9.—Acrydium migratorium (Locust).

(in 1857, in which year a dead locust was also picked up in the Strand in London). The devastations caused by these well known insects have sometimes penetrated to the heart of France. They certainly destroy large quantities of food, but in return they furnish to the inhabitants of the countries to which their visits are most common, excellent repasts. The Arabs, the Egyptians, the Tartars, the inhabitants of Barbary, etc., relish these locusts as much as the Greeks enjoy their *Cicada;* hence locusts are always to be found for sale in the market-places of these people. Indeed cart-loads of them are brought to Fez as a usual article of food; and the Africans, far from dreading their invasions, look upon a dense cloud of locusts as we should so much bread and butter in the air. They smoke them, or boil them, or salt them, or stew them, or grind them down as corn, and get fat upon them!

The custom has prevailed for many centuries, for Diodores tells us that from this circumstance was derived the denomination of *Acrydophaghi,* or eaters of locusts, given to some Ethiopian tribes.*

Locusts belong to the family of *Orthoptera.*

Cicada, another race of insects belonging to the family of *Hemiptera* (or bugs), were formerly employed as an article of food.

Aristotle, Aristophanes, Athenæus, and Ælian among the ancients, mention *Cicada* as an article of diet. These noisy insects were formerly much relished by the Greeks, but their taste for them appears to have been neglected from some unknown

---

* The camels of the Arabs eat cooked locusts readily; deprived of their heads, legs, and wings, and stewed in butter, they are eaten by the Arabs themselves.

cause. They are still eaten by the American Indians, who boil a species known as *Cicada septemdecim*, which is eaten raw by the natives of New South Wales.

Concerning bugs (*Cimex*), which belong to the same family as *Cicada*, although they abound in some parts of Paris and London, we know of no use whatever that could be made of them! Southey once remarked, " We have not taken animals enough into alliance with us. The more *spiders* there were in the stable the less would the horses suffer from flies. The *fire-fly* (*Elater noctiluca*) should be imported into Spain to destroy mosquitoes. In hot countries a reward should be offered to the man who could discover what insects feed upon fleas."

It is well known that *cockroaches* (*Blatta Americana*) destroy bugs, and when a house is infested with one of these noxious insects, it is rare that the other will be found in the same place. But man himself appears hitherto to be the animal that destroys most fleas.

Many more disgusting insects than those just mentioned are eaten in different parts of the world, but as this work might fall into the hands of people of delicate appetites, I shall pass them over, and refer to Kirby and Spence's manual for a description of the *Pterophagi*, a people of Africa, who chase the game upon their own private property.

Aranea edulis, a large spider, is relished by the natives of New Caledonia—this spider is about an inch long; it is roasted over the fire.

Humboldt has seen Indian children drag from the earth centipedes eighteen inches long (probably *Spirostreptus olivaceus* or *S. indus?*), and more than half an inch broad, and devour them.

The same author also speaks of the *Agautle* of the Mexicans, an aliment formed exclusively of the eggs of certain species of the boat-fly, *Notonecta*. These eggs also contribute to the formation of a certain oolitic rock that is being deposited in the great lakes of Mexico, whence M. Virlet d'Aoust and other geologists conclude that the oolitic strata of the Jura, etc., must have had a similar origin.

The Mexicans consume great quantities of these eggs: they find them strewed by thousands upon the reeds on the banks of the great fresh-water lakes, Texcocco and Chalco. They shake them into a cloth, and set them to dry, after which they are ground like flour, placed in sacks, and sold to the inhabitants, who make with this flour a peculiar kind of cake called *haulté*. The unground eggs are also used to feed chickens, etc.*

---

* M. d'Aoust, on his return from Mexico, gave me some of these eggs in 1858; they are very small, oval and white; but I have not yet submitted them to analysis.

Thomas Gage spoke of this peculiar insect product as early as 1625.

The insects whose eggs are taken to produce this Mexican flour are of three species. Two of these belong to the genus *Corixa* of Geoffroy; the first was described in 1831, by Thomas Say, under the name of *Corixa mercenaria*; the other is looked upon as new, and has been called *C. femorata*. But on the same reeds are observed the eggs of a third insect, a new species of boat-fly, which M. Guérin Menneville has termed *Notonecta unifasciata*; this is a larger insect.

We have heard of a beetle called *Chlænius saponaris* (or *Carabus saponarius* of Olivier), of which soap is made in some parts of Africa. This fact is easily accounted for by the great abundance of this insect and the quantity of grease it contains.

Another beetle, *Calandra granaria*, a dark-brown insect, with a spotted thorax, too well known by the ravages it commits in the granaries of southern Europe, contains both *tannic* and *gallic* acid: an extremely interesting fact, discovered by Mitonart and Bonastre, and confirmed by the further researches of Bonastre and Henry. Tannic acid and gallic acid can be extracted from this beetle by means of ether, alcohol, or water. The solution precipitates gelatine and forms ink with salts of iron, etc., characteristic properties of the substances in question.

Fire-flies (*Elater*), of which I have spoken at length in my work on *Phosphorescence*, are employed in some countries as lights, as ornaments, and to kill mosquitoes.

A dipterous insect, belonging to the genus *Stomoxys*, has been spoken of by the Abbé Moigno, formerly editor of the "Cosmos," a French periodical, as capable of producing truffles, hence it has been termed *mouche trufigène*, or the truffle-producing fly. But this subject, which was brought forward by M. Ravel, is an illusion: the persons alluded to thinking that the *truffle* is the product of this fly as the gall-nut is produced by the *Cynips!* It required the entire weight of M. Dufour's evidence to refute these errors, and to convince those concerned that the truffle is a *fungus* like the mushroom, springing from seeds, and not the result of an insect's bite upon the oak-roots. That eminent naturalist showed also that several insects lived upon truffles, and were we to attribute the formation and growth of this fungus to an insect, there are some hundreds which we might look to with equal reason.

I now turn to the common house-fly (*Musca domestica*). Though this insect is not directly useful to us, it contributes, indirectly, to our comforts more than many of us suppose. It is true that Ugo Foscolo used to call flies "one of his three miseries of life," yet the larvæ of these insects nourish themselves upon animal matters which if not disposed of in this manner, would putrefy and evolve noxious gases into the air we breathe; thus the fly doubtless tends to purify the air by preventing the formation of *miasma*.

In this manner, *Musca domestica*, *M. carnaria*, and *M. Cæsar* have their uses. Some flies (the Blue-bottle, etc.), as I have already stated, give birth to larvæ already hatched; others (*M. Cæsar*, etc.) lay millions of eggs, whence proceed, in a day or two, innumerable devourers of dead flesh. One single female of *M. carnaria* (Blue-bottle) will give birth to 200,000 young already hatched; and Redi formerly ascertained that these grubs will devour so much food in twenty-four hours as to increase, in this short period, two hundred times in weight.

This will account, perhaps, for the assertion made by Linnæus, that three individuals of Latreille's *Musca vomitaria* will devour a dead horse as quickly as a lion could do it.

Many beetles devour dead flesh as eagerly as do the larvæ of flies. Stagnant waters are purified by the larvæ of the *Ephemera* flies, etc.

Before quitting the subject of flies, I will mention some curious results obtained lately by M. Berard, who has been studying the influence of light upon animal growth. His observations are applicable to the whole tribe of insects. It appears from them that differently coloured light, or, in other terms, the different rays of the solar spectrum, have a very different influence upon the development of young animals, on the hatching of eggs of insects, the growth of larvæ, etc.

Many philosophers, from the time of Priestley and Ingenhouz to the present day, have studied the influence of light on vegetables, but few have paid attention to its action upon the animal organism. Thus, whilst Priestley, Ingenhouz, Sennebier, De Candolle, Carradori, Knight, Payer, Macaire, and some others, made manifest the action of light upon vegetable respiration, absorption, exhalation, etc.; in a word, upon the phenomena of nutrition and development in plants; Edwards and Morren were almost the only observers who studied animal life from the same point of view. Edwards showed that without light the eggs of frogs cannot be developed, and that the metamorphosis of tadpoles into frogs cannot be effected in absolute darkness.* Again, Moleschott has recently shown that the respiration of frogs is most active in the daylight, diminishing considerably during the night; and Charles Morren observed *Infusoria* to evolve oxygen whilst basking in the sunbeams which play upon the stagnant waters they inhabit.

* Compare Higginbottam in "Proceedings of the Royal Society," 1862; where some experiments of Edwards are refuted.

Later still, M. Berard took a certain quantity of eggs of the fly (*Musca Cæsar*); he divided them into separate groups, and placed them under different coloured glass jars. In four or five days, the *larvæ* produced under the *blue* and *violet* coloured jars were much larger and more fully developed than the others: those hatched under the *green* jar were the smallest. The blue and violet jars were found, therefore, to be most favourable to rapid and complete development; then came the *red, yellow,* and *white* (transparent) jars; and last of all the *green*.

The *larvæ* developed in a given time under the influence of *violet* light were more than three times as large as those hatched and reared in *green* light.*

The experiments are certainly very interesting in a practical point of view; for if it be true, as it appears to be, that the larger a silkworm is the more silk it will produce, it would be worth while to repeat these experiments upon silkworms, and endeavour to raise a large breed under violet glass.

* The effects of the sun's rays, when filtered through differently coloured glass, upon the development of infusorial life, has recently occupied Mr. Samuelson. He fitted up a box containing three compartments, covered by a pane of *blue, red,* and *yellow* glass respectively, and found that under the *blue* and *red* glass infusoria were rapidly developed, whilst under the *yellow* hardly any signs of life were visible. He then transferred a portion of the infusion from the *yellow* to the *blue* compartment, when *infusoria* very soon made their appearance.

Nothing would be easier than to select a portion of some silkworm establishment for the experiment, and to furnish this section of the building with violet-coloured windows. It would indeed be interesting to see these *violet*-coloured panes become as necessary to the silk breeders as the *yellow* window is essential to the photographer. In the former instance the violet would serve to allow the chemical rays of light to pass, while the other rays are excluded. In the latter, the yellow is used to cut off these chemical rays, and to let pass the remainder.

Chapter VI.

Crustacea.

Artificial Propagation practicable with Crustacea as with Fish—The Common Lobster—Laws of Regeneration—The Crawfish—Curious Discoveries relating to the Young of these Animals—Phyllosoma—Zoëa—Metamorphosis among Crustacea—Praniza and Anceus—Larvæ of Lobsters—The Colouring Matter of Lobsters, Crawfish, etc.—Composition of a Lobster-shell—Shrimps—Crangon vulgaris—C. boreas, Sabinea septemcarinata, and other Shrimps—Prawns—Palemon carcinus and P. jamaicensis—Other Species of Prawns—Bopyrus crangorum—The Isopoda—The Family of Crabs—Cancer pagurus—C. mœnas—Pinnotheres—Pagurus—"Diogenes"—Land-crabs—Thelphusa fluviatilis—Crabs of the genus Gecarcinus—Their wonderful Emigrations—Birgus latro, or the Robber Crab—Quantity of fat it produces—Concluding remarks on this Family.

CRUSTACEA.

I NOW leave the useful Insect world to speak of some Crustacea, a class of animals extremely remarkable, both in a scientific point of view and in a practical sense. Lobsters, crawfish, crabs, shrimps, etc., will here demand our attention, and will furnish us many occasions of relating curious or novel details concerning this section of the animal world.

It has lately been ascertained that artificial fecundation and breeding can be effected with some of these Crustacea, as easily as with fish. Messrs. Coste, Haxo, Chabot, etc., have, of late years, devoted much attention to this subject.

A capital of about five shillings, we are told, is sufficient to start with, and, if the business is well managed, the investment will not be regretted. The eggs of a female lobster are taken and placed in a water-trough, and the seed of the male strewed over them; they are then carefully attended to, and nourished upon such substances as observation or experiment prescribes. That is the fundamental principle of rearing Crustacea (Fig. 10).

By breeding crawfish in this manner, some interesting facts relating to the earlier phases of their life have been brought to light.

The common lobster (*Astacus marinus*) is abundant on the rocky coasts of England, and may be seen in clear water, at no great depth, at the time it deposits its eggs, that is, about the middle of summer. It produces from 15,000 to 20,000 eggs. Dr. Baster actually counted 12,444 eggs under the tail of one female lobster, exclusively of those that still remained unprotruded in the body.

The craw-fish (*Astacus fluviatilis*) produces upwards of 100,000 eggs, a fact which has doubtless contributed to the success of the undertakings alluded to above, and which seems calculated to facilitate the artificial multiplication of this species.

Large lobsters are very voracious animals, devouring sometimes their own young, and fighting fearful battles among themselves. When in these skirmishes they lose a claw it soon grows again, but never so large as the lost one it replaces. This power of reproduction of lost parts is extremely developed in lower animals, where the principle of vitality is not concentrated so much in central organs; it is observed to a wonderful extent in

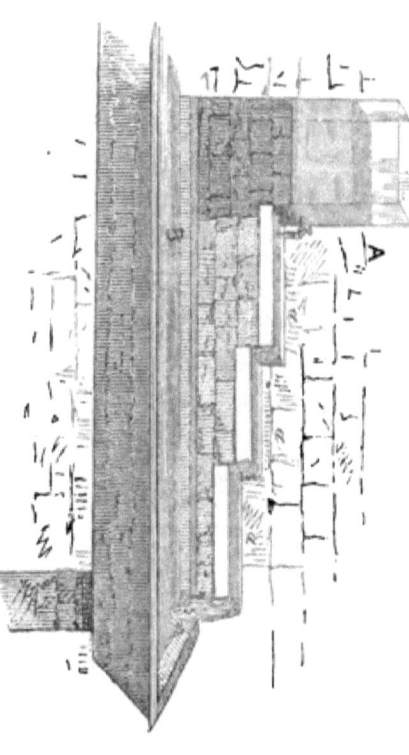

FIG. 10. BREEDING TROUGHS FOR HATCHING EGGS OF CRUSTACEA, ETC. (from a sketch taken at the College de France, Paris).

A. Cistern. *a. a. a.* Glass troughs, containing gravel; the water flows constantly from one to the other in a gentle stream. B. Large trough for salmon, etc.

polyps, sea-anemones, worms, snails, lobsters, lizards, and even in some fish.

Lobsters, in common with most crustaceans, possess the faculty of reproduction to a great extent: if a claw be torn off it is renewed, and if injured the animal will sometimes throw it off of his own accord.* Any violent shock to the nervous system will likewise cause this. Hence, if a lobster be thrown into boiling water or spirits of wine, etc., it will frequently throw off its large claws. Pennant observed that lobsters are apt to cast off their claws during a loud clap of thunder, or by the noise of a large cannon. When a man-of-war meets with a lobster-boat, a jocular threat is used, that if the master does not sell them good fish, the ship's crew will *salute* him!

M. Jobart de Lamballe showed, not long since, that the regenerative force of which we speak decreases as the animal organism becomes more complicated. Hence, if you cut a polyp into two, three, four—one hundred pieces, each fragment will become a new animal. But if we go a step higher—from polyps to worms, for instance — it will be found that, on dividing a worm in two *longitudinally*, the animal will not survive the operation; but if the worm be divided *transversely*, each

---

* See Reaumur, " Sur la Reproduction des Jambes de l'Ecrevisse." (Mem. de l'Acad. des Sciences, Paris, 1712.)

section becomes a new worm. Ascending still higher—to lobsters and fish, for instance—the exterior parts of the body can alone be thus regenerated; and Spallanzani has shown that when the tails of lizards—a class still higher—are cut off, the new tail does not always possess the whole number of vertebral bones; in other terms, the regeneration is incomplete. In animals with warm blood, this regenerative faculty is greatly diminished, but *still exists*, even in man himself. But the same force which in man forms the scar of a wound, or heals the stump after amputation, will with lizards reproduce a tail, with lobsters a claw, with polyps the *whole body!*

The mouth of the lobster, like that of insects, " opens," says Buffon, " the long way of the body, not crossways, as in man. It is furnished with two teeth; but as these are not sufficient, it has three more in its stomach." The latter were formerly used in medicine under the pompous names of *oculi cancrorum*, the *yeux d'ecrevisses* of the French, instead of carbonate of magnesia. The lobster sheds its shell, in all probability once in a year, and then retires under a rock or into a hole until the new skin is again covered with a solid crust. Whilst thus deprived of its hard covering, the lobster becomes an easy prey to most of the inhabitants of the deep, and even to his own species;

so that incredible numbers perish annually, from this circumstance alone, upon our coasts. Under water these curious creatures run swiftly upon their feet, and when alarmed spring from twenty to thirty feet as rapidly as a bird can fly. They are commonly taken in the night by means of a wicker-basket or net, into which a bait, consisting of pieces of flesh or the entrails of fish, has been thrown. The places in which these nets or baskets are lowered into the water are marked by floating buoys.

Very young lobsters seek refuge in the clefts of rocks, and in holes or crevices at the bottom of the sea. There, without seeming to take any food, they grow large in a few weeks' time, being nourished upon the various matters which the water washes into their retreats. When their shell is completely formed, they become bolder, leave the rocks, and creep along the bottom in search of prey. They live chiefly upon the spawn of fish, the smaller crustacea, marine worms, etc. All these facts must be borne in mind by those who undertake to rear them artificially.

The crawfish (*Astacus fluviatilis*) is found in the fresh waters of Europe and Northern Asia. There is a species which inhabits the Mediterranean, and attains more than a yard in length. This is, perhaps, the creature that Aristotle calls ἀστακὸς in his

History of Animals. The common crawfish thrives best in rivers, in holes in the banks, and under stones, where it awaits the small mollusca, fishes, larvæ of insects, and other animal matters, upon which it feeds. The curious old writer, Jerome Cardan, tells us that this animal is a sign of the goodness of the water in which it is boiled, for the best water turns it very red, an absurd notion, like many emanating from this and other similar writers on medicine and natural history in the dark ages of superstition.

Desmarest assures us that a crawfish will live for twenty years or more, and that it becomes larger in proportion to its age. Towards the end of spring it casts off the pieces which form its shell, but in the course of a few days becomes again covered with a solid coating as hard as the previous one, and one-fifth larger. Sometimes this moulting takes place at the end of summer; it appears to depend entirely upon the locality the animal lives in, as it is seen to occur at different seasons in different localities. Its eggs are carried for some time under the abdomen, like those of the lobster. The crawfish is taken in various manners, either by nets or bundles of thorns, in which flesh in a state of decomposition is placed, or by inserting the hand into the holes it inhabits.

By rearing these Crustacea artificially, M. Gerbe,

who was aiding M. Coste in his experiments, discovered that the curious little beings known as *Phyllosoma* are nothing more than the *larvæ* or young forms of the crawfish. The egg of the crawfish, on quitting the mother, becomes a *Phyllosoma*, which is afterwards changed into a **perfect crawfish**. The metamorphosis is as complete as with insects.

Professor Thomson, of Belfast, discovered formerly that certain crabs gave birth to curious-looking beings, to which a French naturalist had previously given the name of *Zoëa*. These *Zoëa*, which were looked upon as distinct animals, turn out to be the *larvæ* or young of other well-known Crustacea. Similar facts have recently been made known by Mr. Couch, of Penzance.* But since the publication of Professor Thomson's observations, we have, in the order of *Entomostraca*, examples of generation equal to that we mentioned in speaking of the *Aphides* in a preceding chapter. M. Hasse has also shown that the curious creatures known as *Praniza* are only *larvæ* of *Anceus*, so that metamorphosis is doubtless as active in Crustaceans as in Insects.

It is now an established fact, therefore, that the eggs of crawfish bring forth *larvæ* which do not resemble the parent, but were formerly classed as

* Brit. Ass. Report, 1857.

distinct animals, under the name of *Phyllosoma*, and that crabs' eggs produce *larvæ* known formerly as *Zoëa*. Moreover, it has lately been shown by Valenciennes that lobsters produce *larvæ* also, and that these were also taken for *Zoëa*.

In the year 1853, M. Etienne Leguilloux sent to the *Jardin des Plantes* of Paris some young lobsters barely hatched from the eggs. It was soon discovered that these young creatures were the identical Crustaceans formerly described by M. Bosc as *Zoëa*. After a space of eight days, these *larvæ* change their skins or moult for the first time; at two months old their change of form becomes very evident; at the age of three months the large claws which characterize the lobster begin to show themselves, and at six months old the transformation is complete. These creatures have then the form of the adult lobster. In this state they are often caught on the shore, and sent to the French markets under the name of *Quatre-quarts*. They fetch a much higher price, in proportion to their size, than the full-grown lobster.

The black or dark-blue colour of lobsters and their allies is very remarkable, in a chemical point of view, as it becomes *red* in hot water. Macaire and Lassaigne have examined its nature, but little is yet known of it. In its natural state it is a very dark bluish-green fatty matter, which becomes red when exposed to a heat of 70° (centigrade), and in this state resembles the red colouring matter extracted by Goebel from the legs and beaks of certain geese and pigeons. It can be extracted from the lobster's shell by means of alcohol, in which it is soluble; but during the operation the colour turns red. Sulphuric and nitric acids turn the red alcoholic solution to a permanent *green*, which the alkalis do not again change to red. This is one of its most remarkable properties. A permanent organic *green* is such a desideratum at this moment in the tinctorial world, that the discovery of a new dye of that description would be worth thousands of pounds!

Moreover, the red colour of the lobster can be modified by chemical means; for instance, with oxide of lead it produces a violet combination, and the dark-coloured shell becomes red when it is put in contact with acids, alkalis, certain salts, etc. It also turns red by long exposure to the air, by putrefaction, etc.; but it does not change colour in carbonic acid gas, or in hydrogen. Chlorine bleaches it completely.

The hard envelope of Crustacea is formed principally of carbonate of lime, a little phosphate of lime, and a few other salts in small proportions. All these are intimately mixed with a certain amount of animal tissue.

Shrimps resemble lobsters and crawfish to a certain extent; they have been subdivided by naturalists into many distinct groups.

The *Crangon vulgaris* is our common shrimp, which, according to Pennant, is the most delicious of all Crustaceans.

In the Arctic Seas we have two other descriptions of shrimps, namely, *C. boreas* and *Sabinia septemcarinata*, which are sometimes plentiful on the west coast of Davis's Straits.

Other species of shrimps are found on the coasts of Mexico, in the Mediterranean, the Indian Ocean, etc., so that this tribe of Crustacea is pretty widely diffused.

Besides shrimps, we have also numerous species of prawns, shrimp-like Crustaceans belonging to the genus *Palemon*, well-known to the epicure. Some varieties found in hot climates attain one foot in length: such are *Palemon carcinus* of the Indian Seas and the Ganges, and *P. jamaicensis* of the Antilles.

Prawns generally inhabit sandy bottoms near the coasts, but are often found at the mouths of rivers, even far up the stream, at some distance from the sea.

The common prawn of our markets is *P. serratus*. It is taken on the English, Flemish, and French coasts, where it is accompanied by two other species,

P. squilla and *P. varians*, which both differ a little from the former.

There is a kind of shrimp belonging probably also to the genus *Palemon*, and which is about seven inches long; it is very common at the mouths of rivers in Florida. Leba has called it *the American crawfish*, but it is probably the *Palemon setiferus* (Olivier) of naturalists.

Shrimps and their allies are the principal *scavengers of the ocean;* they clear away the decomposing animal matter which floats in the sea. They are highly prized as a delicious and nutritive article of food, and might be easily reared artificially or cultivated, as crawfish and lobsters have been in France, were it deemed profitable or necessary.

Curious little parasitical Crustacea belonging to Latreille's genus *Bopyrus* are found living upon prawns. Those who are in the habit of eating prawns will probably have sometimes observed a tumour under the carapace on one side of the animal. On lifting this part of the shell, the parasite will be discovered immediately under it, upon the branchiæ or gills. These little beings belong to the family of *Isopoda*. The species which live on our common prawn is *Bopyrus crangorum*. The former does not appear to suffer at all from the invasion of this parasite, which will one day, doubtless, turn out to be the *larvæ* of some other

Crustacean—perhaps of the prawn itself. Be that as it may, the section of *Isopoda* presents a wide field of experimental research, from the wood-louse, *Oniscus murarius*, which used to enter into the composition of certain quack pills, upwards.

Let us now turn to the family of crabs. Our large edible crab (*Cancer pagurus*, L.) is taken upon the *rocky* coasts of Great Britain, Ireland, and Western Europe; it is rarely met with on sandy coasts, such as the littoral of Flanders. Pennant says that it casts its shell every year between Christmas and Easter; but Lyell, in his "Principles of Geology," says that a crab taken in April, 1832, on the English coast, had its shell covered with oysters of six years' growth; hence it was concluded that this crab could not have moulted for six years.

Like other Crustacea, it is probable that the crab moults once a year in its younger days, but it has not been ascertained at what period this moulting ceases.

As to artificial breeding and rearing, I shall refer to what has been said of lobsters and crawfish.

Cancer mænas, L., is a much smaller and less-esteemed edible crab, common on our coasts. A still smaller species is the pea crab (*Pinnotheres pisum*), which is about the size of a spider; it is found sometimes, in the month of November, living in the interior of the shells of mussels. Other small species inhabit the shells of other living mollusca.

The Hermit Crab (*Pagurus Bernardus*), an indigenous representant of a numerous and interesting group, is not sought for as food in this country. Being deprived of a shell of its own, it inhabits the shells of large univalve mollusca (*Buccinum undulatum*). There are many species of *Pagurus* that live in holes at a considerable distance from the sea, which they only visit now and then, as we go to our watering-places. Thus the hermit crabs of the far west come to the sea once a year, to lay their eggs and change their shells. Some of them are eaten by the native Americans, but they sometimes disagree with strangers. Catesby says that a species known as "Diogenes," found at the Antilles in the shell of a large periwinkle (*Turbo pica*), is roasted in this shell by the natives, and esteemed delicate eating. Though the whole body of the *Pagurus* is soft and tender, its anterior claws, which project from the shell it inhabits, are so strong, that an individual of two or three inches long pinches smartly. When some of these species are taken they emit a feeble cry,* and endeavour to seize the enemy with their strong claws.

* The production of sounds by aquatic animals is rare. On sounds produced by fish, see Dufossé in "Comptes Rendus," Paris Academy, 1858, and again in the same publication for 1861.

But some of the most useful and most remarkable of crabs are undoubtedly the *land crabs*, which belong to the genera *Thelphusa* and *Gecarcinus*. Of the former some live far away from the ocean, under damp stones in the woods; others, such as *T. fluviatilis* (Fig. 11), which would be taken by a

FIG. 11.—Thelphusa fluviatilis (European land-crab).

casual observer for a small common crab, burrows in the earth on the banks of rivers. This animal is about two and a half inches long, and of a yellowish colour; it was known to Hippocrates and Aristotle, and is represented on certain ancient medals. The Greek monks eat it raw, and the Italians feed upon it during Easter. It is not uncommon in the south of Italy, Greece, Egypt, and Syria.

The crabs of the genus *Gecarcinus* resemble that just mentioned. They abound in the hilly districts of the Antilles, where they are known to the French as *Toulourous*. They are likewise found in the tropical parts of America, Asia, and Africa. During the day they hide themselves in damp holes or cavities of trees and rocks, or lie motionless under damp blocks of stone. Although, like fish and other Crustacea, etc., they are furnished with branchiæ or gills for breathing, they cannot live in the water. At certain periods of the year, generally about the month of May, they unite in troops, and make long excursions over the country towards the sea, where they repair to lay their eggs. Thus once a year they march down to the sea-beach, some thousands at a time, laying waste everything they meet on the road. They proceed in so direct a line, that no geometrician could send them to their destination by a shorter course. They travel by night and repose by day, unless it happen to rain, when they profit by the circumstance, and proceed by day also.

On arriving at the sea-shore, their eggs are deposited in the water, and the mother crabs, leaving accident to bring them to maturity, wander back to their accustomed haunts. About two-thirds of these eggs are immediately devoured by shoals of fish, brought, as it were by instinct, at this particular time to the shore. The young *Gecarcini* that escape are hatched upon the sand, and soon after millions of these little creatures are seen quitting the shore, and slowly travelling up to the woody mountains.

These crabs are sometimes called *Violet crabs*. They live upon leaves, rotten wood, fruits, etc. They are considered delicious food in the countries where they abound, especially during the time of moulting. In the Carribbee Islands they form a very important element of nutrition.

The elegant writer, Bernardin de St. Pierre, in his "Etudes de la Nature," speaks of these land crabs thus:—

"Il y a des animaux qui ne voyagent que la nuit. Des millions de crabes descendent aux Antilles des montagnes à la clarté de la lune en faisant sonner leurs tenailles,* et offrent aux Caraïbes, sur les grèves stériles de leurs îles, leurs écailles remplies de moelles exquises."

The *Birgus latro,* or robber crab (Fig. 12), is another terrestrial species, and is sought for as food in certain countries. It is remarkable for the manner in which it climbs trees, to feed upon their fruit. The crabs of this species bore a hole at the feet of trees in Amboyna and other islands in the South Pacific Ocean. The naturalist Herbst appears

---

* Buffon says, that "to intimidate their enemies, they often make a clattering noise with their claws during their march." Their nippers are very strong, and a crab of this species loses its claw rather than let go its grasp. One of them may be often seen making off, having left its claw still holding fast upon the enemy. The faithful claw seems to perform its duty to the utmost for upwards of a minute after its owner has retired.

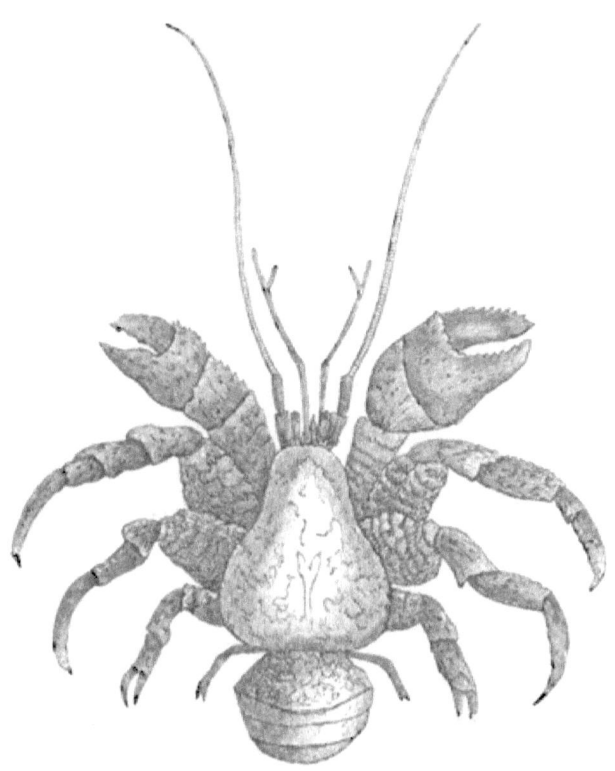

Fig. 12.—Birgus latro (Robber Crab—individual capable of producing one quart of oil).

to be the first who studied this remarkable crab, and to his accounts we are referred by Rumphius, Seba, Linnæus, and Cuvier. The Indians say that these robber crabs can live upon cocoa-nuts, and that they make their excursions during the night. Quoy and Gaimard have fed them for months upon cocoa-nuts alone. They climb principally a species of palm-tree (*Pandanus odoratissimus*), and devour the small palm-nut that grows thereon. They are a favourite article of food among the natives. Darwin observed the *Birgus latro* in the Keeling or Cocos Islands, situated in the Indian Ocean, about six hundred miles from the coast of Sumatra. He assures us this crab grows to a monstrous size. M. Liesk tells us he has seen the *Birgus latro* open cocoa-nuts, which they perform, according to Darwin, by tearing off the exterior fibres or husk, and then striking them repeatedly upon the "eye-holes," with their heavy claws.

The young are hatched and live for some time on the shore. The adult *Birgus* proceed at times to the sea to moisten their gills; the journey is made at night. They make their beds of cocoa-nut husks. These crabs are not only very good to eat, but under the abdomen of the larger ones is lodged a mass of fat, which, when melted, yields *as much as a quart of oil;* so that a native having such an animal at his disposal can make his supper of the crab, and light himself to bed with the oil. It would be interesting to examine this oil, and ascertain the quantity that could be produced annually by a given number of these crabs.

* * * * *

The Crustacea of which we have spoken, and whose study we now relinquish, are all oviparous, and have separate sexes; therefore artificial breeding and cultivation of any of their species would probably be attended with success. The artificial breeding of crawfish and lobsters appears to have begun in France; M. Coste of Paris, and M. Gaillon of Concarneau, have lately concentrated their attention upon the artificial propagation of these and some other useful animals upon the French coasts.

Chapter VII.

Mollusca.

CEPHALOPODA.

India and China Ink—Fossil ink-bags—Octopus vulgaris—The colour "Sepia"—Sepia officinalis, or "Cuttlefish"—Cuttle-bone—Loligo vulgaris—Edible Cuttlefish—Chemical nature of their Colour—Nautilus—Argonauta—Carinaria.

GASTEROPODA.

The Tyrian purple—Curious properties of the colouring matter of Sea-snails—Murex brandaris—Purpura lapillus—Helix fragilis—Yandina fragilis—Purpura patella—Murex truncatus—Experiments with American Sea-snails—Colour furnished by Whelks—Buccinum—Influence of light upon the production of their colour—Process used by the ancients to dye purple—Uric acid in Gasteropoda—Murexide—Snails that are reared for food, etc.—Helix pomatia—Snail gardens—H. aspersa—H. horticola—Arion rufus—Chemical Analysis of Snails—Limacine—Helicine—Uric acid in H. pomatia—Turbo littoreus, or Periwinkle—Haliotis—Snails used as money—Cypræa moneta—Other species of Cypræa—"Love-shells"—

Conus—Oliva—Ovula—Strombus gigas—Cassis—Turbinella—Murex—Buccinum—Curious experiments with Snails—Slugs—Limax maximus—L. agrestis.

BIVALVE-MOLLUSCA.

Mytilus edulis, or common Mussel—Its culture, etc.—Hurtful at certain seasons—M. choros—M. Magellanicus—M. arca—M. lithophagus—Ostrea edulis, or common Oyster—Details concerning its artificial breeding and propagation, etc.—Acclimatisation of Mollusca—Fishing on the Plessix bed—Spondylus—Cardium edule, or Cockle—Solen—Pecten maximus—Tellina—Tridacna gigas—Chama—Cameos—Stone Cameos and Shell Cameos—Chinese Cameos—Pearl-oysters—Avicula margaritifera—A. frimbriata—A. sterna—Pearl Fishery—Its extent—Pearls of Mytillus edulis—Anodontes—Unio pictorum—Unio margaritiferus—Culture of the Fresh-water Pearl-Mussel—Artificial modes of causing it to produce pearls—Pinna—Their silky byssus and its uses—Their pearls—Other uses of shells—Tunicata and Bryozoa.

MOLLUSCA.

ON the first order of Mollusca, that of *Cephalopoda*, we meet with many animals both curious and useful. These singular creatures, among which the common Cuttlefish may be taken as an example, derive their name, *Cephalopoda*, from the fact that their feet seem to be placed upon their head. Their body is fleshy and soft, generally somewhat cylindrical; their head is distinct from the body, and is furnished with particularly large eyes; their mouth, placed at the top of the head, has two strong horny mandibles something like the beak of a parrot, and is surrounded by long fleshy tentacles or arms (often termed *feet*), which are almost always provided with numerous suckers, by means of which the animal grasps tightly anything that comes in its way. Indeed, so firmly can the *Cephalopoda* adhere to foreign bodies by means of these suckers, that it is easier to tear away the arm or tentacle than to release it from its grasp; but the animal, on the contrary, can release itself instantaneously, as numerous observations show. They walk upon the bottom of the ocean, head downwards, making use of their tentacles as feet.

The different varieties of Cuttlefish are provided with a very peculiar organ, generally known as "*the ink-bag*"—a purse-like sac filled with a dark-coloured liquid, which is secreted by a special gland. When the animal is irritated or frightened, it empties a quantity of this fluid into the water to conceal itself.

This coloured liquid was used by the ancients as a kind of ink, and it has been affirmed that it formed the basis of several paints, among others of China or *India ink;* but the latter often owes its colour to the charcoal of burnt cork, or to common lampblack mixed with glue.

The drawings with which Cuvier illustrated his studies of the *Sepia, Loligo,* and other *Cephalopoda,* were executed with the ink furnished by the animals he was dissecting.

Miss Mary Anning, of Lyme Regis, formerly discovered that the ink-bags of certain fossil *Cephalopoda* in the Lias beds has been preserved unaltered to the present day, though it must have lain buried in the strata for myriads of centuries! "In the lower Jura formations" (the lias of Lyme Regis), says Humboldt, "the ink-bag of the *Sepia* has been so wonderfully preserved that the material which, myriads of years ago, might have served the animal to conceal itself from his enemies, still yields the colour with which its image may be drawn."

After this, my discovery that the fossil *Teredo* of the Brussels Tertiary formations have a powerful odour of the sea, when freshly taken from the earth and broken, is less astonishing.*

Certain *Cephalopoda* swim or dart about more or less swiftly in the water, and have even been seen to leap out of the sea like the flying-fish. This is observed with certain species of *Loligo*, or "*Pen-fish.*"

Octopus vulgaris (*Sepia octopodia*, L.) has eight tentacles, furnished with double rows of suckers. It is common enough in the European seas, and in summer destroys great numbers of lobsters on the coasts of France. It is from this species that the brown colour called "Sepia" was formerly extracted. It is known in English as the *Eight-armed Cuttle* or *Poulp*, and when it attaches itself to the arms or legs of a bather is very difficult to get rid of, though they are generally timid creatures, and only fight as a last resource.

The common Cuttlefish (*Sepia officinalis*), whose shell or bone is often thrown upon our coasts by the waves, is probably well known to our readers.

* "Comptes Rendus of the Acad. des Sc.," Paris, July, 1856.

Its bone, which supports the soft parts of the animal's body, is employed to polish ivory and bone objects, to prepare tooth-powder, and for a host of minor uses. It is known in the shops as "Cuttle-bone," or when powdered as "Pounce." It is frequently hung in the cages of Canary birds, who clean and sharpen their beaks by pecking at it. This bone exists in other animals of this group: in *Loligo vulgaris* (the common Calamary) it is almost transparent, and sloped somewhat like a pen, whence this and other allied species are sometimes called Pen-fish. *Loligo vulgaris* is common on our coasts. The colour of its almost transparent greenish body changes at intervals, and adapts itself to that of the water it inhabits. In all the so-called naked* *Cephalopoda* the colour of the skin is highly changeable, showing spots which brighten and fade with a rapidity superior to the cuticular changes of the chameleon; a faculty which they owe to a very remarkable cuticular tissue, which has often engaged the attention of anatomists.

Hardly any sea is without some species of naked *Cephalopoda*; their food consists principally of fish and crustacea, but they are very voracious, and will devour almost any kind of animal matter. Their flesh, especially that of the tentacles, is edible, and

* To distinguish them from those possessed of shells (*Nautilus*, etc.)

is considered nutritious. They are not eaten in Britain, but in other countries the Cuttlefish is sometimes sought as food. In the Neapolitan market-places, for instance, the arms or tentacles, cut into portions and prepared for cooking, are to be frequently seen. They resemble the lobster in flavour. According to Aristotle, they were esteemed as food by the ancients, and the old writer Athenæus informs us how to prepare a cuttlefish sausage.

Prout, Bixio, and Kemp have examined the colouring matter produced by these animals, and contained in their ink-bag. It appears from their researches to be very similar in nature to the black pigment of the eye of other animals. It is insoluble in water, but remains for a very long time suspended in the liquid, as we observe with finely pulverized chalk. This principle is known to chemists as *Melaïne*.

About 12 cwt. of *cuttle-bone* (of *Sepia officinalis*, L.) arrives yearly in Liverpool; it is mostly sold to druggists, who use it chiefly for making tooth-powder. The dried contents of the ink-bag is imported from China to Liverpool, at the rate of a few pounds annually. It either arrives in cakes or is made into cakes, called *Sepia* and *Indian ink*. Imitation Indian ink is made of cork charcoal, soot, etc., as I have already observed.

Besides these naked *Cephalopoda*, there are some which possess very splendid shells: such are the

Nautilus and the beautiful *Argonauta*, or Paper Nautilus, which is not unfrequently seen, on calm days, gliding softly on the surface of the blue Mediterranean, and of which Pliny, Buffon, and others have given such poetical descriptions. Their shells are sought for as ornaments. Other species, such as certain rare *Carinaria*, produce magnificent shells, which sell at a high price for drawing-room ornaments.

The *Nautilus pompilius*, according to some naturalists, is seen floating on the waters of the Atlantic between the tropics; the *Argonauta Argo* on the Mediterranean; the *Carinaria fragilis* also inhabits the Atlantic; whilst *C. vitrea*, a rare species, is chiefly found in the South Seas.

* * * * *

In the second order of Mollusca, named *Gasteropoda*, we have some very interesting, useful, and ornamental animals. To save space and time required for minute description, the common Garden Snail or Slug may be taken as an example of the order of *Gasteropoda*. The species of this large tribe are very numerous, and perhaps as beautiful or as useful as numerous.

I shall mention, in the first place, the *Gasteropoda* from which the ancients extracted the colouring matter known as *Tyrian purple*. This magnificent colour, only worn by kings and nobles, was the produce of a sea-snail.

Many rather marvellous tales have been related concerning the origin of this purple dye of the ancients. At the present time, all that appears to be known with certainty is, that its discovery was made at Tyre, and that it was produced by certain sea-snails. Some writers assure us that the species which furnished the colour were *Murex brandaris* and *Purpura lapillus* (Fig. 13); of which the first

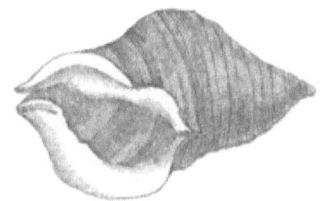

Fig. 13.—Purpura lapillus (Purple-producing Whelk).

produced the finest and most expensive colour, and the latter, which is as common on the English coasts as upon those of the Mediterranean, is a kind of *whelk*.

The liquid which can be squeezed out of this whelk is *colourless*, or nearly so; but by the action of light it becomes first of a *citron* tint, then *pale green, emerald green, azure, red*, and finally, in about forty-eight hours, a magnificent *purple*. To enable the colouring matter to take successively all these tints, it must not be allowed to dry.

At the meeting of the Jerusalem Literary Society, held November 14, 1857, Dr. Roth, of Munich, gave the results of his researches upon the ancient Tyrian purple dye. He shows that in the works of Pliny and Aristotle the names of *Buccinum*, *Murex*, and *Conchylia* are so vaguely used, that nothing on this subject can be learned from them. Hasselquist, according to Dr. Roth, supposes the true shells to be *Helix fragilis*, L., and *Yandina fragilis*, the mollusca of which are purple, and stain the fingers; but their dye is not lasting. When Dr. Roth first came to Palestine, he found at Jaffa the *Purpura patula*, the snail of which is sought by the native Christians as food during the fast-days. On puncturing this animal there issued a greenish liquid, which, when exposed to the sunshine, changed to purple. This purple increased in brilliancy when it was washed. Comparing this with the accounts left by the ancients, Dr. Roth thinks the colour he produced is evidently their blue-purple, for they had a blue-purple, a deep-purple, and a red-purple.

Between Soor and Saida, according to the same author, the *Murex truncatus*, or *trunculus*, is found in abundance, and its colour is more brilliant than that of the *Purpura*. One of these *Murex* is sufficient to dye a square inch of cloth, which would require five individuals of *Purpura patula*. Wool takes the dye better than any other substance; silk takes it with difficulty. Dr. Roth appears to have assured himself that the liquid extracted from these snails becomes coloured under the influence of light, and that the air has nothing to do with it; but I fancy both agents are active. The eggs of these sea-snails are laid in June, and hang upon the rocks in large balls. They have also a purple colour.

Researches similar to those just mentioned have been made before. Long ago, Thomas Gage reported that certain shells found near Nicoya, a little Spanish town of South America, possessed all the dyeing properties noticed by Pliny and other old writers. They were employed for dyeing cotton on the coast of Guayaquil and Guatemala. In 1686, Cole made similar observations on the English coasts. Plumier formerly discovered a colouring snail in the Antilles, and Reaumur made repeated experiments on common whelks (*Buccinum*), which he picked up on the coast of Poitou. Duhamel repeated these experiments on *Purpura*, found in abundance on the shores of Provence. He and Reaumur first noticed the extraordinary influence of light in the production of the colour. Bixio studied, though incompletely, the colour furnished by *Murex brandaris*, and found it to be identical in properties with that furnished by other *gasteropod* mollusca.

The art of dyeing purple was continued in the

East as late as the eleventh century, at which epoch it still existed in all its vigour. The process employed and the manner of taking the snails has been described by an eye-witness, Eudocia Macrembolitissa, daughter of the Emperor Constantine VIII. Her book is to be found in the first volume of the collection published in 1781 by M. d'Anse de Villoison, entitled "Anecdota Græca," etc. The process was as follows:—A quantity of *Gasteropoda* were pounded in a trough, and to the mass thus produced was added either a quantity of urine in a state of putrefaction, or some water in which a certain number of the pounded snails had undergone putrefaction. The cloth was soaked in the liquor produced by this mixture, and acquired a purple colour on being exposed to the air; sometimes it was warmed a little, to accelerate the production of the colour.

Jacobson and De Blainville found *uric acid* in these snails, as a product of the so-called *saccus calcareous*, an organ which secretes uric acid in snails and other Mollusca.* Now, Dr. Prout formerly transformed uric acid into a purple colour of great beauty, which he termed *purpurate of ammonia*, and which Liebig has since called *Murexide*. It appears

---

* This organ is supposed to be the first vestige of a kidney. See Jacobson in "Journ. de Phys.," xci. 318; and compare Carus, "Comp. Anat.," tom. i. p. 377, fr. ed.

evident at the present day that this substance derived from uric acid is identical with the purple of the ancients. Dr. Sacc has used it as a dye very recently, and obtained tolerably good results; and Dr. Schlumberger has endeavoured to prove that the varied hues of parrots, humming-birds, pheasants, etc., are owed in great measure to murexide. At the present time, large quantities of murexide have been obtained from guano, which contains much uric acid, for the purpose of dyeing. It is a splendid substance when pure, presenting in one direction beautiful metallic green reflections, and in others brown and purple tints.

But to this we must add, that, up to the present time, no rigorous chemical experiments have been made with the purple colouring matter extracted from sea snails, and the curious manner in which it is developed under the influence of the sun's rays seems to indicate that it is really distinct from murexide, however much the latter may resemble it.

Many snails are sought for and bred as articles of food or medicine. Among the terrestrial species, *Helix pomatia*, or the Apple snail (Fig. 14), known in France as the *Grand escargot*, is cultivated to a considerable extent, and is eaten, principally during Lent, in France, Belgium, Germany, and other parts of Europe. Indeed, the taste for this animal has so much increased lately, that the oyster trade suffered last year in France, in consequence of the number of these snails brought into the markets.

These land snails shut themselves up for the winter in a curious manner, by means of what is

FIG. 14.—Helix pomatia (Edible Snail).

called an *operculum*, a flat circular piece of shell-like substance, just large enough to cover the opening of the animal's shell, to which it is attached by a strong mucous cement. The snail, having previously fixed itself to a wall or a tree by means of the same glutinous substance, or buried itself among the dead leaves, remains throughout the winter in this state, without food, until the warmth and moisture of spring recalls it to life.

In countries where snails are used as food, they are only taken whilst in this state of hybernation. They are reared and fattened in what are called *snail-gardens* (*escargotoires*, French).

A snail-garden consists either of a large square plot of ground boarded in, the floor of which is covered half a foot deep with herbs, or of broad shallow pits sunk in the ground. In these the snails are kept. **They are** fed with fresh leaves, bran, and potatoes during summer; **and in winter, when they fix** themselves against **the walls of the pit, they are collected, packed in casks, and sent to market** (see fig. 15, p. 153).

Four millions of snails are sent annually from the snail-gardens of the town of Ulm, in Germany; and this is no monopoly, for the other snail-gardens of Germany are in a flourishing state.

Helix pomatia is not so common in England as on the Continent; it is found abundantly, however, near Dorking. Some naturalists believe **it to have been** accidentally introduced into England, at a comparatively **recent period;** but others suppose it to be **indigenous to the British Isles**, though rare. I **have frequently observed** very fine specimens in the neighbourhood of Brussels, where **the climate seems** to suit it remarkably, and where its **cultivation** would doubtless succeed admirably.

Helix aspersa, our common Garden Snail, **is not** deemed worth the trouble of cultivation, so long as the former larger species can be obtained. **It is** distributed over a large portion of the globe; we **find it, or at least varieties of it,** at the foot of

Chimborazo, in the forests of Guiana and Brazil, and on the coasts of the Mediterranean in Europe, Asia, and Africa, as well as in the British Isles, Belgium, Germany, etc.

The latter species, as *H. pomatia*, *H. horticola*, etc., when boiled in milk, is said to afford a light and strengthening food for invalids; and for many years the large Apple Snail (*H. pomatia*), the Red Arion (*Arion rufus*)—a reddish-brown slug, often met with in damp places, and extremely common in the neighbourhood of Brussels—and a few others, have been employed in medicine, in the form of sweet syrups, for colds, sore throats, etc. Their emollient qualities are owing to the large proportion of mucilage they contain. Braconnot extracted 8 per cent. of this mucilage and 84 per cent. of water from snails; the remainder consisted of a few substances not well known, the principal of which he has called *limacine*.

M. Figuier says that alcohol extracts from *H. pomatia* a medicinal substance, which he calls *helicine*, although it appears to be a mixture of different principles, the nature of which has not been determined, and, in all probability, does not differ from the substance called "helicine" by Dr. De Lamarre of Paris, who has employed it for many years in the treatment of phthisis. It is, however, but another of the thousand and one pharmaceutical secrets, and if it have any advantage over most of the others, it is that it contains nothing hurtful or poisonous.

M. Mylius, unaware of the discovery of Jacobson mentioned above, has found *uric acid* in *H. pomatia* immediately between the shell and the animal, whence it can be extracted by water. By shaking the snail in water, the uric acid is separated, and soon deposits itself, as an insoluble powder, at the bottom of the mucilaginous liquid thus produced.

Among sea snails, the common Periwinkle (*Turbo littoreus*), one of the most common Mollusca in our latitudes, and small Whelks (*Buccinum*), which are eaten with a pin, together with several of their allies, are extensively used as food. The heaps of periwinkle shells that are seen at the outskirts of fishing villages on the coasts of England, Belgium, etc., suggest that some use ought to be made of them. In soils which are deficient of lime, these shells might be coarsely powdered, and spread over the ground.

A species of *Haliotis*, sometimes called the Ear-shell, a large, handsome *Gasteropod*, whose shells, when polished, present the most varied and magnificent tints, with mother-of-pearl lustre, and which are easily recognized by the circular holes perforated along the edges of the shell, is frequently seen in the shops for sale as an ornament.

In *Haliotis tricostalis* (*H. padollus* of other authors) the shell is furrowed parallel with the line of perforations. *H. tuberculata* may be taken as a type of these curious Mollusca. There are seventy-five species of *Haliotis*, which are scattered widely over the world. A species that abounds on the coasts of the Channel Islands, where it goes by the name of *Omer*, is cooked, after being well beaten to make it tender; other species are eaten in Japan. The shell of the larger specimens, taken in the warmer parts of the ocean, is much used for inlaying and other ornamental purposes, for which it is very valuable.

We must not imagine that the breeding or cultivation of snails is a modern undertaking, for Varro, in his "De re Rusticâ," speaks of the enormous size to which snails may be brought by culture. Pliny, in his Natural History, repeats Varro's statements, and says that the large species of snail was a favourite dish with the Romans, who were in the habit of breeding and fattening them in snail gardens, similar to those now seen on the European Continent (Fig. 15).

A certain number of *Gasteropoda* are sought after for the beauty of their shells. The *Cowries*, certain species of *Cypræa*, are still used as money by the Africans, the natives of the Laccadives, and other Indian islands. The *Cowrie*, properly so called,

FIG. 15.—Snail Garden near Ulm, Germany.

Cypræa moneta, L., has been imported into Liverpool of late years at the following rate:—

In the year 1851, 1704 cwt. of *Cypræa moneta*; in 1852, 2793 cwt.; in 1853, 1680 cwt.; in 1854, 90 cwt.; in 1855, 311 cwt. There are two commercial varieties of White Cowrie—one called the Live Cowrie, taken when the animal is alive in the shell; the other called the Dead Cowrie. Both are largely collected in the Maldive Islands, and exported to Africa, where they are used as money, and exchanged for palm-oil, ivory, gum, etc. They are found upon the shores of the warmer seas, principally in the Mediterranean and Indian Seas.

Other species of *Cypræa*, known to the French as *Porcelaines*, or as *Pucilages*, and by the English as "*Love-shells*," are used as ornaments, etc. Children sometimes place them to the ear, to listen, as they say, to the sound of the sea.* The small *Cypræa* are made into clasps, buttons, ear-rings, bracelets, etc. (Fig. 16), and even into stags, elephants, horses, etc., for children. They are not only hawked about the streets in England, but exposed for sale in the shop-windows of Continental

---

* The peculiar noise that is heard when one of these shells, or indeed any object of a somewhat similar shape, is placed to the ear, has never been clearly explained. It appears, however, to be owing to the movement of the air in and out of the shell, the current being caused by approaching the cold shell to the ear.

sea-ports, where they are entitled "*Animaux en Coquilles à* 1 *fr.* 25 *c.*"

The larger species of *Cypræa* were consecrated by the Greeks at Cnidos, in the temple of Venus.

Fig. 16.—Ancient Egyptian Necklace of Love-shells (Cypræa), ornamented with Gold.

In certain parts of Africa the natives worship them as idols, or, at least, used to do so a few years ago. In more civilized countries, superstitious people wore them as a talisman, to protect themselves from certain maladies.

Almost all the species of this genus inhabit the warmer parts of the Atlantic, the Pacific, and the Mediterranean. A very small species is found on our coasts.

The large spotted shells belonging to the *Gasteropod* genus *Conus,* or Cone, on account of the shape of these shells, and those of the genus *Oliva,* are seen as ornaments on the chimney-piece. Their price is somewhat high.

The Mollusca belonging to the genera *Cypræa,*

Oliva, Ovula, etc., sometimes quit their old shells, and produce new ones.

The Conch-shell, the product of *Strombus gigas,* is much prized as an ornament when the aperture is of a fine rose colour. This large shell is a common chimney-piece ornament, but it is also used for making cameos; and the inferior kinds are purchased also by the masters of potteries as a source of pure lime, or for other purposes. Great numbers are sold for ornament. It is taken principally on the shores of the West Indies, and is imported from time to time into Liverpool, at the rate of from 6000 to 11,000 shells per annum.

The allied Mollusc, *Cassis* (or Helmet shell), is sometimes preferred for cutting cameos. *Cassis rufa* is exported from the Maldives to Italy for this purpose in considerable quantities.

Certain species of *Murex* and *Buccinum* are also purchased as decorative ornaments.*

The *Gasteropod* known as *Turbinella pyrum* (or *Voluta gravis,* Linn., Fig. 17), produces a large pear-shaped shell, which is much prized in India for making bracelets and other ornaments. This shell has acquired a certain commercial importance, and

* Most of the shells mentioned in this work are to be seen in the collection at the British Museum, and many have been elaborately drawn and coloured in Lovell Reeve's extensive work on Mollusca, in 20 vols.

is commonly called "the Chank-shell." They are fished for on the coasts of Ceylon, in the Gulf of Manaar, on the coast of Coromandel, etc., where they are brought up by divers from depths of two to three fathoms of water. Those taken with the snail inside are most esteemed; the dead shell, thrown upon the beach by the tide, having lost its

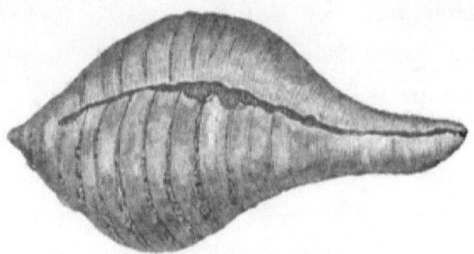

FIG. 17.—Turbinella pyrum (Chank-shell).

enamel, is of little value. The number of these shells imported at Madras from Ceylon is quite astonishing. In the year 1854, 1,875,053 *Turbinella* shells arrived there to supply the manufacturers of ornaments; in 1858, 1,268,892 shells were imported; and in 1859, 1,910,050. Indeed, the Chank fishery at Ceylon formerly employed six hundred divers, and yielded a revenue of £4000 sterling per annum for licences. It is now free. Sometimes 4,500,000 of Chank-shells are obtained in one year in the Gulf of Manaar, valued at upwards of £10,000 sterling.

The principal demand for these shells is for

making *bangles*, or armlets and anklets, the manufacture of which is almost confined to Dacca. The solid porcellanous shell is sliced into segments of circles, or narrow rings of various sizes, by a rude semicircular saw. The *bangles* thus constructed are worn by the Hindoo women; they are beautifully coloured, gilded, and often ornamented with precious stones (Fig. 18).

These same *Turbinella* shells are also used frequently as oil-vessels in the Indian temples, for which purpose they are carved and ornamented.

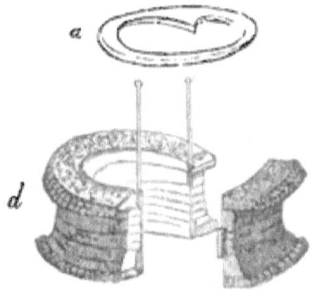

FIG. 18.—Hindoo Bangle, made from the Chank-shell.
a. Segment of the shell. *d*. Segments united to form a bangle or bracelet.

In Dacca, on account of its weight and smoothness, the shell of *Turbinella pyrum* is used for calendering or glazing, and in Nepal for giving a polished surface to paper.

The value of these shells imported in the rough state into Madras and Calcutta, from the 30th of April, 1851, to the 30th of April, 1859, is represented for Madras as £34,184, and for Calcutta, £29,985.*

Sir Emerson Tennant has given an account of this shell, under the name of *Turbinella rapa*.

In a preceding chapter I mentioned the curious manner in which lost or mutilated organs are regenerated or replaced in inferior animals, and even in some of the higher classes. This regenerative faculty is very remarkable in snails, and Mollusca in general. When a snail's shell is broken, the animal repairs it in an astonishing manner; and when some part of the animal's body has been cut away, it also reappears. Spallanzani, having cut off a snail's horn, observed that it began to bud out again in about five and twenty days, and continued to grow until it was as long as the other. He then cut away part of the head of another snail, and in course of time the lost portion was renewed. When the head was cut completely off the experiment sometimes failed, and the animal died; but more than once a new head grew again even in this case; at the end of a few months the snail appeared with another head, in every respect similar to the lost one. The snails thus operated upon retired into their shells the moment decapitation had taken place, and covering the opening with their *operculum*, remained thus enclosed for weeks, and even

* See "The Technologist," vol. ii. (1862), p. 185.

months. When forced out for examination at the end of thirty or forty days, some appeared without any marks of renewal; but in others, especially when the weather was warm, a fleshy globule, of a greyish colour, was observed about the middle of the trunk.

No particular organization was noticed in this globule, but in eight or ten days it became larger—rudiments of lips, mouth, tongue, and the smaller horns appeared, then gradually developed, and in the course of two or three months the injury was so completely repaired, that the new head could only be distinguished from the old one by its lighter colour.

These experiments have been confirmed by Bonnet, Schœffer, Gerordi, and others.

Snails have been divided into two genera, in one of which (*Slugs*) the animals have no shell. The large slug (*Limax maximus*, L.), whose body is grey spotted with black, is frequently seen in damp cellars, gardens, etc.; and the small slug (*L. agrestis*, L.), after summer showers, in kitchen gardens. These have not yet been turned to much account by man; on the contrary. But the red slug (*Arion rufus*, L.) is still used in country places for cough mixtures, etc.

The *Snails*, properly so called, belong to the genus *Helix*. Of them I have spoken at length;

their species can often be determined by the form and colour of their shells.

* * * * *

I shall now turn to the Bivalve Mollusca, as examples of which the Oyster and the Mussel may be taken.

The common mussel (*Mytilus edulis*), which lives in the sea, and is quite distinct from the fresh-water mussel, of which I shall speak further on, is found on our coasts in considerable quantities, and also upon the rocky coasts of almost the whole of Europe. These mussels live fixed to the rocks or piles, to which they attach themselves by means of their *byssus*, a sort of silky hair which the animal secretes for this purpose. In some genera allied to mussels, such as the *Pinna* of the Mediterranean, this byssus attains a foot and a half in length, and the inhabitants of Palermo sometimes use it to make gloves and stockings. Its chemical nature does not appear to have been examined.

At certain seasons mussels are extensively consumed as an article of food, for which purpose they have been actively cultivated. For many years they have been bred artificially in salt-water marshes that are periodically overflowed by the tide, the fishermen throwing them in at the proper seasons. The animals, being undisturbed by the agitation of the sea, and protected from the inhabitants of the deep, cast their spawn, and multiply wonderfully. It was soon found that it required only one year to people a mussel-bed of considerable size, and that one-tenth may be left to renew the bed completely after the harvest.

The mussels are taken from these beds from July to October, and, though sold at a moderate price, their commerce is not without importance, many thousands of these mollusca being annually dispatched from the coasts into the interior.

After it had been discovered that a breed of oysters might be crossed with other breeds, and produce new varieties of oysters, similar experiments were attempted with mussels, and have met with considerable success, especially in Italy, and in the Bay of Aisguillon, in France.*

It has been found that the mussels, which live suspended to piles, ropes of vessels, nets, etc., attain to a much greater size than those which live on the bottom, whether this be sandy, rocky, or muddy. This fact has been turned to advantage by the Italian and French mussel-breeders; thick ropes, suspended to wooden piles, are placed in the water of the mussel-beds, as represented in the engraving; the mussels adhere to these ropes by their byssus, and the ropes are then tightened

* D'Orbigny's "Hist. des Parcs à Moules de l'Arrondissement de la Rochelle," La Rochelle, 1847 ; and De Quatrefage's "Souv. d'un Nat.," tome ii. p. 360, *et seq.*

a little, so that the animals no longer lie upon the bottom, but live suspended in the water (Fig. 19).

Mussels are apt to become very hurtful as food at certain seasons of the year, from May till the end of August, a period denominated by the French "*la période des mois sans r.*"

The cause of this does not appear to be satisfactorily ascertained. Some attribute it to the presence of spawn in their gills during this period;

Fig. 19.—Breeding Mussels upon ropes, as practised at La Rochelle, France.

others assert that mussels become unwholesome from having eaten the spawn of the common starfish. The latter casts its spawn precisely from the beginning of May till the end of August. However, the fact does not appear proved. In cases of indisposition from this cause, small doses of ether, frequently administered, have proved beneficial.

The genus *Mytilus* is pretty numerous in species, most of which are used as food in different countries. *Mytilus choros* is a large mussel, seven or eight inches long, found on the coasts of the island of Chiloe, on those of South America, etc. The animal is as large as a goose's egg, and is said to be of a fine flavour. There is another variety still larger. The natives cook them in the following manner:—A hole is dug in the earth, in which large smooth stones are placed; upon these stones a fire is made, and when they are sufficiently heated, the ashes are cleared away, the mussels are heaped upon the stones, and covered over first with leaves and straw, then with earth, and left to stew. This appears, from certain accounts, to be not only an ingenious, but very superior mode of cooking mollusca.

In our *Mytilus edulis* small pearls are frequently found—I shall have something to say on pearls presently—and in the month of November the small Pea-crab (*Pinnotheria*) is often seen in their shells.

Mytilus Magellanicus, which inhabits the southern coast of South America, is a mussel four or five inches long, whose flesh is well flavoured and nutritious. Its shell is easily recognized by its longitudinal furrows.

Other species, such as *Mytilus arca* of my friend

Professor Kickx (that Van Beneden calls *Dreissena polymorpha*, and which has been honoured with a host of other names besides), are probably carried about the world on the keels of ships, and very widely diffused.

The species just mentioned, *M. arca*, is found inhabiting seas, lakes, rivers, marshes, etc., extending over nearly the whole surface of Europe, from lat. 43° N. to lat. 56° N. It is, moreover, found in the earth in a fossil state.*

A highly-nutritious mussel, *Mytilus lithophagus*, L. (or *Modiola lithophaga*, Lam.), common enough in the Mediterranean and at the Antilles, has the faculty of burying itself alive, as it were, by penetrating into wood, stones, and rocks, as the *Teredo* and *Pholas* bore into ships.

The *M. lithophagus* form, even in the hardest rocks, cavities which they can never leave, in consequence of their increasing in size as they grow older.

The common oyster (*Ostrea edulis*), a bivalve mollusca, too well known to need description here, is subject to great variation. Many different varieties have been observed in nature, or artificially produced by culture. A single oyster brings forth from one to two million of young, of which the

* On this curious mussel, see Van Beneden in "Ann. des Sc. Nat., 1835."

greater part perish before achieving their development, if they are abandoned to themselves in the ocean.

These animals spawn about the commencement of spring, and, according to most naturalists, they fecundate their own eggs;* but instead of abandoning its spawn, like many other shell-fish, the oyster keeps it lodged between the gills, where it undergoes the process of incubation. This process coutinues for some time, and that is why oysters are not generally esteemed from May to September.

But the depth of the water in which the oyster lives seems to have a considerable influence upon the time of spawning. In its first state, the young oyster exhibits two semi-orbicular films of transparent shell, which are constantly opening and closing at regular intervals. As they grow larger they attach themselves to the rocks ; but for this purpose they do not secrete long silky strings, as the mussels do. When they find nothing solid to adhere to, they become cemented together in large quantities, each adhering to its neighbour, and constitute solid shoals or oyster-beds, which sometimes

* The gasteropod and bivalve mollusca are all hermaphrodite ; but with the snails and slugs we have been studying, the concourse of two individuals (four organs) is necessary to ensure reproduction ; with bivalves, such as the oyster, it appears the male organ can render fertile the products of the female organ in the same animal.

attain many leagues in length and a considerable thickness. Leuwenhoek counted upwards of three thousand young oysters moving about in the liquid confined in the interior of the valves of the parent mollusc. These minute beings are provided with shells in about twenty-four hours after the eggs that produced them are hatched.

M. Gaillon says that the oyster feeds chiefly upon a green animalcule, called *Vibrio navicularis*; but others assert that it lives also upon vegetable substances, such as the mucilage of sea-weeds, etc.

The liquid contained in oyster shells has a composition very different from that of sea-water; it contains a notable amount of albumen, besides numerous animalculæ and flocculent vegetable matter. It has lately been analysed by Payen, who finds it composed of 85·98 parts of water, 1·33 of organic matter, and 2·85 of mineral salts and silica. Ether has the property of coagulating and throwing down the albumen contained in this liquid.

Some varieties of oyster live attached to the roots or branches of trees that are periodically covered by the rising tide. At the mouths of rivers in South America and other tropical countries, groups of magnificent oysters are seen thus suspended together with that curious bivalve, *Perna ephippium*, and are rocked to and fro by the balmy sea-breeze when the tide retires. These are called mangrove oysters, as they hang chiefly upon the root-like branches of the mangrove (*Rhizophora mangle*), which propagates itself in an extraordinary manner along the muddy banks of tropical rivers.

Oysters which live suspended in this manner grow to a much larger size than those which lie in shoals at the bottom of the sea, as we observed was the case with mussels. At St. Domingo the negroes cut them off with a hatchet, and they are served upon the table with the roots.

Oysters have been cultivated more or less for centuries; the ancients attached great importance to this great cultivation. The Romans cooked them in a great variety of manners; and Apicius, a glutton who lived in the time of Trajan, is said to have possessed a peculiar secret for fattening oysters. Britain has been celebrated for its oysters since the time of Juvenal. Pliny informs us that Sergius Orata got much credit for his stews of Lucrine oysters, "for the British oyster was not then known." Among the antiquities discovered at Cirencester, a Roman *oyster-knife* was found, and presented to the British Association in 1856.

The art of propagating these mollusca in artificial oyster-beds has been much perfected of late years. The works of M. Coste, who has studied this question *in extenso* on the borders of the Mediterranean and on the coasts of the Atlantic, will be consulted with profit by all oyster-breeders.

On the western coast of France, where the water is somewhat deep, it was found that the oyster requires *five* years to arrive at its complete growth, whilst in shallow water *two* years are amply sufficient.

A model plan for breeding oysters may be seen in the lake of Fusaro, in Italy, where mussels and oysters are *cultivated* with much success—*where almost the entire quantity of spawn is developed without loss.* That oysters can be transported from one coast to another, and that oyster-beds can be artificially produced on coasts which are deprived of them, was proved by an Englishman more than a hundred years ago.

Guided by this knowledge and his own researches, M. Coste lately proposed to the French Government to form a chain of oyster-beds all along the western coasts of France. Several beds exist there at present, but most of them are falling to decay, and others are completely exhausted. M. Coste has already commenced operations. He gets fresh oysters for propagation from the open sea; he turns to advantage those that are rejected by the trade; and, lastly, he collects the myriads of embryo oysters which, at each spawning season, issue from the valves of the oyster, and which are

Fig. 20.—Bundle of Faggots for Propagating Oysters, according to M. Coste's system.

now lost to commerce for want of some contrivance to prevent their escape and inevitable destruction.

Every oyster, I have stated, produces from one to two million of young; out of these not more than ten or twelve attach themselves to their parent's shell; all the rest are dispersed, perish in the mud, or are devoured by fish! Now, if bundles made of the branches of trees, faggots of brushwood, or any similar objects, be let down and secured to the oyster banks by weights, the young oysters will, on issuing from the parent's valves, attach themselves to these faggots, and may, on attaining perfect growth, be taken up with the branches, and transported to places where it is desirable to establish new oyster-beds.*

I witnessed the success of this experiment made upon the coast of Brittany, not very long ago. If the process of transportation take place at the proper period, success is almost certain. Between the months of March and April, 1858, about 3,000,000 oysters, taken from different parts of the sea, were distributed in ten longitudinal beds in the Bay of St. Brieuc, on the coast of Brittany. The bottom was previously covered with old oyster-shells, and boughs of trees arranged in bundles.

* I called attention to some of these facts (which I consider of importance to oyster-breeders), on December 7, 1861, in an English periodical.

To these the young oysters attach themselves; and so fruitful were the results, that one of the *fascines* that was examined at the expiration of six months, was found to have no less than 20,000 young oysters upon it (Fig. 20).

A report furnished to the French Government shows that about *twenty-five thousand acres* of coast may be brought into full bearing in three years, at an annual expense not exceeding £400.

But to ensure the continuous propagation of artificially-formed oyster-beds, the dredging must be effected at proper intervals.* For this purpose the beds must be divided into zones, and one-third of each zone only be dredged each season. In this manner an absolute repose of two years is allowed to each of the zones.

Hitherto, the dredging used to take place in September, the spawning season being then over; but in that very month the young oysters attach themselves to their parents' shells, so that the mollusca are disturbed at a moment when the new population is beginning to form. To avoid this, M. Coste has proposed to fix the dredging season in February or March.

In England there have been many Acts of Parliament passed for the protection of oyster-

* Dredging is performed with a strong net, having an iron rod at its base.

beds. The fisheries are at present, however, regulated by a convention entered into between the English and French Governments, and an Act (6 and 7 Vict. c. 79) passed to carry the same into effect, which enacts that the fisheries shall open on the 1st of September, and close on the 30th of April.

It has been said that the Romans formerly discovered that different varieties of oysters could be intermixed so as to produce cross-breeds superior in every respect to the stocks whence they sprang. Of late years, a medical man of Morlaix, in France, took some of those large unpalatable oysters termed *pied-de-cheval*, and crossed them with some small Ostend oysters. The result exceeded his expectations, and he produced a new breed of large oysters, equal in delicacy to the small ones of Ostend.

The Ostend oysters, which are in such high repute in Belgium, are fished upon the English coast, and bred in artificial oyster-beds at Ostend.

Mr. Robert Macpherson, speaking of the common oyster, says :—" The *Ostrea edulis* of Linnæus is subject to much variation, which has occasioned the making of one or two questionable species, and rendered uncertain the limits of its distribution. The common English and Welsh oyster is, however, certainly abundant and of excellent quality at Redondela, at the head of Vigo Bay; and I have likewise dredged it off Cape Trafalgar in sand, and off Malaga in mud, but have not noticed it further eastward in the Mediterranean."

It is a curious fact that oysters become sooner developed in shallow water, and are then by far the most highly-esteemed for the table. Moreover, oysters that are dredged in deep water far from the coast expel from their shell the whole of the water it contains, the moment they are taken from their natural element; whilst those which are taken on the coast, from beds which are daily deprived of water by the retiring tide, preserve the water contained in the valves of their shells, and can be transported to great distances without losing their freshness. Thus the American oyster, one of the many varieties of *Ostrea edulis*, is imported alive into Liverpool at the average rate of sixty-five bushels a year.

In November, 1861, the French papers *Le Journal du Havre* and the *Moniteur*, announced the success of an experiment, made with a view of acclimatizing American mollusca on the French coast. M. de Broca, M. Coste, and Count de Ferussac, took part in the undertaking, and on the coast at Hogue Saint Wast breeding-beds were prepared. In 1861, the steward of the "Arago" steamer brought over about 200 oysters, and the same quantity of *clams*, a shell-fish consumed in great quantities in the United

States. These were deposited in the beds of Saint-Wast, under M. Coste's immediate superintendence, and in November following it was ascertained that the specimens were healthy, and promise to supply abundance of spawn for the propagation of the species on all the coasts of France. This experiment has induced M. Coste to make preparations for acclimatizing on the French littoral all the best kinds of mollusca from different parts of the globe, and we learn that Professor Agassiz has offered his aid in this useful undertaking.

The opening of the oyster fisheries at the mouth of the river Auray, in France, coincided on the 30th of September 1861, with the meeting of the Agricultural Society of the province, presided over by the Princess Bacciocchi. At two o'clock in the afternoon, 220 fishing-boats, covered with flags and flowers of all descriptions, sailed out to the oyster-beds, in presence of an immense concourse of people, which had spread itself over the bridges, along the quays, on the side of the mountain Du Loch, and all along the port of Auray, the weather being magnificent. The boats anchored on the Plessix bed, about half a mile from the port, and commenced dredging. *In the short space of one hour the product of this fishing amounted to* 350,000 *oysters.* In the evening the little town of Auray was illuminated, and dancing kept up out of doors to a late hour by the peasants and the fishermen. It is the first time that the culture of the oyster has been thus brilliantly inaugurated. Some days after this little *fête*, 320 fishing-boats, carrying 1200 men, began dredging off the same beds. *Twenty millions of oysters* had been brought into port when I commenced this chapter.

Among oysters, a genus of mollusca called *Spondylus* are remarkable for their curious shells, which are covered with long spines; there are about twenty-five species of them, inhabiting the warmer parts of the ocean, the Mediterranean, etc. They are collected as curiosities. A host of useful bivalves, belonging all to this immense family of Lamellibranchiate Mollusca, to which the oysters and mussel belong, crowd upon us.

To begin with the least important of them; every one knows the common *Cockle* (*Cardium edule*). The genus *Cardium* is very widely distributed. The species are generally found buried in the sand on the sea-shore. Many of them attain a considerable size. Our common cockle forms an abundant and nutritious article of food, especially in seaport towns.

The curious mollusca belonging to the genus *Solen*, or Razor-shell, are frequently picked up on our coasts. They furnish us an example of a bivalve shell which is many times wider than long (though an ordinary observer would say it was much longer than wide). On the coasts of Scotland, where the specimens are very fine, they constitute an article of food.

Pecten maximus, or the common Scallop, frequently met with on our coasts, is also an edible species, and, when properly cooked, is considered a delicacy. Other species of Pecten, more beautiful, are sought as ornaments, and employed as such in different ways. I have seen elegant ladies' purses constructed with these shells. In the same manner are the pretty little pink and yellow shells of the *Tellina* (common enough on some of our coasts), utilized in the shops to construct various kinds of ornaments, to decorate workboxes, pincushions, etc.

The largest shell known is that of the immense oyster, *Tridacna gigas*, which inhabits the Indian seas. It is known in English as the Clamp-shell; the French term it *bénitier*, because one of its valves resembles the fount which contains the holy-water (Fig. 21) in Roman Catholic churches.* The smaller

FIG. 21.—Tridacna gigas (The *Bénitier* shell of the French).

---

* The two holy-water founts (*bénitiers*) in the church of St. Sulpice, Paris, are valves of the *Tridacna*. They were presented by the Venetians to François I. A friend of mine has an elegant ornament for cards, letters, etc.: in the place of the wooden cross (Fig. 21), is a statuette of Venus rising from the sea.

specimens are indeed sold in considerable numbers attached to crucifixes made to hang against the wall. This shell is also sought for to manufacture knife-handles, penholders, and a number of elegant ornaments of various descriptions.

To the same group belong the shells of the genus *Chama*, which attain also a considerable size. These and the shells of the Gasteropoda, *Strombus* and *Cassis*, mentioned before, are those with which *cameos* are made.

Real or stone cameos are cut at great expense in certain varieties of onyx, agate, or jasper. The art of cutting these hard stones is very ancient, and the ornaments thus produced realize a very high price, especially when the workmanship is of a superior quality. They are still cut in Italy, principally at Rome; but cameo artists are not unfrequently met with in other parts of Europe.

The practice of working cameos on shells, and producing what is called a *shell cameo*, has been introduced at a comparatively modern period into Italy. It is carried on to a great extent at Rome in the present day. Shell cameos are much easier to execute than *stone cameos*; hence, however beautiful the design, they are much less valuable than the latter. A good stone cameo, the size of half-a-crown, with a simple head as device, is frequently worth a thousand francs (£40); whilst a shell cameo of the same description, unless of extraordinary merit, would rarely fetch fifty francs (£2).

Cameos are executed on shells as on stones; the subject is worked in relievo on the white portion or outer crust of the shell, while the inner surface, of a pink or brown tint, is left for the ground. Cameo artists who work upon shells are to be met with in London and Paris. The only shells that I have seen employed are the Conch shell (*Strombus gigas*) and the Helmet shell (*Cassis*) among the Gasteropoda, and the shells of the genus *Chama*. The latter mollusc inhabits the intertropical seas; the species lives fixed to the rocks; and its *foot* (or under part of the body by which the animal moves) is remarkable from being bent, and resembling in form the foot of a man. The species known to the French as the *Came feuilletée* is one of the most curious, and may be taken as a type of the group. The superior valve of the shell is composed of superposed plates or layers of calcareous matter of different colours. The cameos made from it resemble closely those cut upon agate or onyx.

I have seen very beautiful cameos cut in Paris upon the ordinary Conch shell (*Strombus gigas*), and sell at eighty francs (£3 6s.). Probably other shells might be found to answer the same purpose; it is sufficient that they present two or more layers of different colours, which is not unfrequently the case with some of the larger Gasteropoda and Bivalves of the Southern seas.

There exists a peculiar kind of cameo termed the *Chinese cameo,* or *pearl cameos.* The process by which they are made has lately been discovered:—

"The Ningpo river abounds in oysters, which the natives take up when they have grown to a certain size. The shells are then partially opened, care being taken not to injure the animal, and moulds bearing the required design are introduced between the valves. The shell is then allowed to close, and the oysters thus operated upon are placed in beds prepared for their reception. After remaining there for some months, they are again taken up and opened, when the mould is found beautifully crusted over with mother-of-pearl; it is then dexterously detached, and made into various ornaments."

We will now turn our attention to the Mollusca which produce *pearls.* Of pearl "oysters," as they are generally called, or rather *pearl mussels*—for the animals that furnish us with these jewels are more closely allied to the mussel than to the oyster—there are two descriptions, namely, those which inhabit rivers or fresh water, and those which live in the sea.

We shall have to consider, then, the *fresh-water pearl,* and the *marine* or *Oriental pearl;* but as the latter is the most important, I shall speak of it first.

On the shores of those countries where pearl oysters abound, they are sought for as eagerly as we seek for *Ostrea edulis* on our coasts. We have seen how the latter is at present drawing the attention of practical men, who are endeavouring to perfect its breed, and to propagate its species widely. Such will doubtless happen one day for the pearl oyster, whose products are so valuable; for not only does this mollusc produce the pearl—

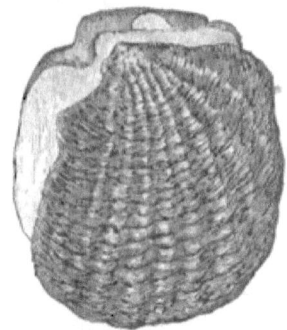

Fig. 22.—Avicula margaritifera (Pearl-oyster).

the "jewel of the sea,"—but also that beautiful substance known as *mother-of-pearl*, with which buttons, knife-handles, penholders, work-boxes, and ornaments of every description, are constantly manufactured.

The animal in question is the *Avicula margaritifera*, L. (Fig. 22). Its shell, of a semicircular form, is of a greenish tint on the outside, and of a beautiful pearly lustre in the interior. It constitutes mother-of-pearl, which is an important article of commerce at the present day. The *pearls* for which this mollusc is also sought are small, accidental excrescences found in the shell, often buried in the animal's body, but most commonly seen adhering to one of the valves of the shell itself. Like other animals of the mussel kind, *Avicula margaritifera* secretes a *byssus*, by which long silken thread it adheres to submarine objects.

Other Mollusca which inhabit the ocean have been observed to produce pearls. Such are the common oyster (*Ostrea*), many mussels (*Mytilus*), and some bivalves belonging to the genus *Perna*. They are also produced by certain fresh-water mussels (*Unio*).

The exact nature of a pearl has been the object of much discussion. Some inquirers imagine it to be the result of a particular disease, which causes the animal to produce these pearly concretions, by occasioning in some parts of the shell an unwonted production of calcareous matter. This being produced abundantly and suddenly, does not spread itself uniformly over the interior surface of the valve of the shell, but constitutes those little concretions we call pearls.

In the opinion of others, pearls are regarded as

a secretion produced by the animal in perfect health, with a view of strengthening certain portions of its shell, either on account of a slight fracture, or to close up apertures pierced in it by marine worms, or, again, to furnish strong points of adherence for certain muscles or ligaments of the animal's body. Be this as it may, Linnæus, in his experiments on fresh-water mussels (*Unio*), discovered a means of causing the mollusc to produce pearls artificially, as we shall see presently.

As to the geographical distribution of *Avicula margaritifera*, which produces mother-of-pearl and the real Oriental pearl, it is found in the Persian Gulf, on the coasts of Arabia Felix, on the coasts of Japan. It is at Cape Comorin, and in the Gulf of Manaar, at the island of Ceylon, that the most productive and celebrated pearl fisheries have been established. Oriental pearls are likewise met with in America, on the coasts of California, at Madagascar, and at the island of Tahiti.

The Gulf of California is about 700 miles long, and from 40 to 120 miles in width. One of the first shells discovered in its waters was a pearl oyster, the *Avicula fimbriata* (*Margariphora mazatlantia* of others), to obtain which the Spaniards, in the seventeenth century, employed from 600 to 800 divers; the value of the pearls obtained amounted annually to about 60,000 dollars. This traffic was so exhausting to the pearl oyster beds, that the fishery is now almost entirely abandoned. Occasionally, however, a shipload of pearl-shell is sent to Liverpool, and sold at the rate of £2 to £4 per cwt. for manufacturing buttons, ornaments in mother-of-pearl, etc.

There is another species of *Avicula*, *A. sterna* of Gould, known to exist in the same locality.

Avicula margaritifera, like other mussels and oysters, lies in banks or beds of greater or less depths. On the west coast of Ceylon these shoals occur about fifteen miles from the shore, where the depth is twelve fathoms; and there, at Aripo, Chilow, Condatchy, etc., the greatest of all pearl fisheries has been carried on for centuries. The season for fishing always commences in March or April, because in those latitudes the sea is then in its calmest state. The fishing continues till the end of May.

The boats of the pearl-fishers hold about twenty men, ten of whom are experienced divers. These descend rapidly through the water to the rocks on which the mollusca are clustered, by placing their feet upon a large stone attached to a rope, the other end of which is fastened to the boat. They carry with them a second rope, the extremity of which is held by two men in the boat, whilst to the other extremity, held by the diver, is fixed a strong net or basket. Every diver is armed with a powerful knife, by means of which he detaches the *Avicula* from the rocks, and which serves to defend him in case he is attacked by a shark. There are marvellous stories told of the length of time these divers can remain under water; but persons who have inhabited Ceylon for many years assure us that they never saw a diver remain submerged for more than *fifty seconds* at a time. They plunge and relieve each other by turns, from daybreak till about ten in the forenoon, when the sea-breeze sets in, and the whole flotilla return to shore. In a short time we shall probably see those iron head-cases and tubes, now used by the divers at work in the Thames, adopted by those of Ceylon. The pearl oysters are taken from the boats, and heaped upon the shore to putrefy. For this purpose an enclosed space of ground is allotted to them. As soon as the putrefaction is sufficiently advanced, the shells are taken and placed in troughs, where sea-water is thrown upon them. When decomposition sets in, the body of the mollusc soon ceases to adhere to the shells and the pearls they contain, which are then taken out, washed, and assorted. The pearl fishery of Ceylon, in 1857, brought in £20,550 15s. 6d.; the same year chank-shells, before mentioned, realized £188 9s.

Such is the present state of things. Our readers

will perceive what a vast field for amelioration is offered here, and what a great improvement it would be to do away not only with the barbarous mode of diving, by breeding the *Avicula* in appropriate places, but with the unwholesome process of extracting the pearls and shells from the putrid heaps of mollusca.

There is no doubt, from the experiments already made with the common oyster, that the pearl oyster might be easily submitted to culture; as it is, the pearl banks in Ceylon, according to Sir Emerson Tennent, were, from 1834 to 1854, an annual charge, instead of producing an income to the colony. *Seven years* is the period required, in the present state of things, before the pearl oyster arrives at perfection, and can be sought with advantage! Diving-bells, or the diving apparatus used in constructing bridges, would be a protection against sharks, etc., though accidents from this cause seldom or ever occur; the noise of the boats seems to scare the sharks away.

According to Dr. Kelaart, the pearl oyster can sever its byssus and change its place, so as to migrate to some distance in search of food, or to escape from impurities in the water, and so moor itself again in more favourable situations. This may account somewhat for their disappearance at intervals, and the bad crops yielded by localities which were abundant in produce the previous season.

In Europe the white pearls are most valued, whilst the inhabitants of Ceylon prefer those of a rose colour, and the Indians and other Asiatic people those which are yellow. Pearls, indeed, vary much in colour and appearance; some are quite black, others dark blue or purple, with a silvery or golden lustre.

During the process of fishing, few places are more lively than the western point of Ceylon. The shells and cleansed pearls are bought and sold on the spot, in small bamboo huts erected for the purpose; and, besides this trade, the confluence of crowds of strangers from different countries attracts dealers in all sorts of merchandize. The long line of huts is a continuously animated bazaar; all is life and activity. But as soon as the fishery closes, scarcely a human being, or even a habitation, can be seen for miles, and the most dreary solitude prevails until the ensuing year.

According to Woodward, the largest pearl known is said to belong to a Mr. Hunt. It measures two inches in length and four inches in circumference, weighing 1800 grains.

The nacreous lustre of the pearl-shell is an optical phenomenon, termed *interference*; it occurs on glass which has lain in the earth for a length of time, and has become decomposed at its surface; the same is likewise seen on the feathers of humming birds, parrots, etc., and in certain chemical preparations.* It is too complicated a subject to be discussed here.

Up to the present time no attempt has been made to cultivate, to propagate artificially, or to acclimatize in other seas, the pearl oyster of Ceylon. To give an idea to what extent the pearl fishery is prosecuted at the present time, I will quote a passage from the " Colombo Observer," (1858), which is as follows :—

" A letter of the 20th March states—' We have had ten days' fishing, and there is about £15,000 already in the chest. There will be ten days' more fishing. Oysters sold to-day as high as twenty-five rupees per thousand."

The shell of *Avicula margaritifera* is imported to Liverpool from the East Indies, Panama, and Manilla, at the average rate of 490 tons per annum. Pearls are frequently imported from the East Indies, but there is no account kept of the quantity.

It is not unusual to find small pearls in the common edible mussel (*Mytilus edulis*), but they are seldom large enough to be of any value. It might, perhaps,

* I have discovered that most substances possess this property, when they are viewed in a proper direction in the sunshine. Polished iron, ebony, and other descriptions of hard wood, possess it to a remarkable degree.

be possible to cause this mussel to manufacture larger pearls. However, such as they are, the pearls of *M. edulis* have been for many years an article of commerce in England.

There are two kinds of fresh-water mussel which resemble each other very closely; the first are found in pools and other stagnant waters, and are known in English as "Pond mussels" (*Anodontes*). The other description inhabit running water, and are seen in sparkling streams. These belong to the genus *Unio*, and are those to which I am about to draw attention.

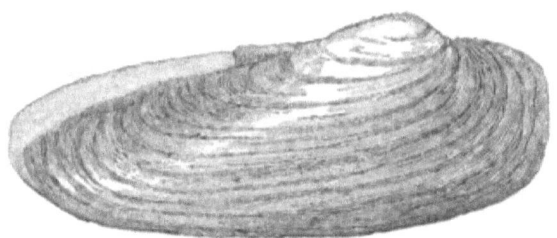

Fig. 23.—Unio margaritiferus (Fresh-water pearl-mussel).

Our readers are probably acquainted with the "painter's mussel" (*Unio pictorum*). It is seen in the shop-windows of vendors of pencils, colours, and engravings, with its edges gilt. It is used by miniature painters to hold colours, and that is all I have to say of it. A much larger and by far more interesting mollusc is the fresh-water pearl mussel (*Unio margaritiferus*) (Fig. 23), a species which is common enough in England, Wales, Scotland, Germany, etc. It has a large bivalve shell, which, when clean, is of a peculiar yellowish-brown colour, with a wide blue band round the edges. The species has been known for ages in Scotland, where it produces pearls (sometimes called "Scotch pearls") that are now and then quite equal to the Oriental pearl of the *Avicula*. Old writers assure us that it was these English jewels that tempted Julius Cæsar to renew his visit to our island.

Unio margaritiferus is as common in Germany as with us. Very fine specimens are seen in the brooks and rivulets of the Bavarian woods and the mountains Fichtelgebirge. Its pearls have likewise attracted attention, and although they are not equal to the Oriental pearl, they are held in certain estimation by the jewellers; and the rich collection of Bavarian pearls that figured some years ago at the Industrial Exhibition of Munich, proved that in Germany the culture of the pearl may one day become a considerable branch of industry. A step has indeed been taken already in this direction. An accomplished geologist, Dr. Von Hessling, of Munich, was directed, a few years back, by the King of Bavaria, to make minute investigations into the manner in which these pearl mussels live, and under what circumstances they produce their jewels, for all the shells do not contain pearls. Dr. Von

Hessling was also directed to examine whether the artificial propagation of *Unio margaritiferus*, with a view of producing pearls, is practicable. The results of his labours were published in 1859 at Leipzic, in an 8vo volume of 376 pages, entitled, "Die Perlmuscheln und ihre Perlen," etc., to which interesting work I refer those who would undertake similar experiments in England.

Two descriptions of pearls are collected and turned to account in Wales. They are known in England as the "Conway river pearls." The first, which are of little value, are taken from the common mussel (*Mytilus edulis*), at the mouth of the river Conway. The others, which are frequently very fine, are taken further up the stream, from the shells of *Unio margaritiferus*. As early as 1693, a paper was published in the "Philosophical Transactions," by Sir Robert Redding, who states that at that period an extensive fishery for these pearls was carried on by the natives who lived near the rivers in the west of Ireland. "Although, by common estimate," says the author, "not above one shell in a hundred may have a pearl, and of those pearls not above one in a hundred be tolerably clear, yet a vast number of fair merchantable pearls, and too good for the apothecary, are offered for sale by those people every summer assize. Some gentlemen make good advantage thereof, and myself saw a pearl bought in Ireland for fifty shillings, that weighed thirty-six carats, and was valued at £40," etc.

In 1842 letters from Norway mentioned that there had been found in the bed of the great stream that runs through Jedderen, in the district of Christiansand, and which, from the excessive heats, became dry, a great number of fresh-water mussels containing pearls, some of which were so fine that they were valued at £60 a piece. At the beginning of the seventeenth century, when Norway was annexed to Denmark, the Government took the pearl-fishery of this stream into its own hands, and the finest pearls were sent to Copenhagen to be deposited in the Crown treasury. After this the produce of the fishery became so low that it did not pay the expenses, and it was abandoned.

Unio margaritiferus is very plentiful in the river Conway, about a mile above the ancient bridge of Llanrwst, near the domain of Gwydir, where the water is beautifully clear, rapid, and deep. It may be taken from this spot up to Bettws-y-Coed.*

I will terminate what I have to say of these pearls by a word upon their *artificial production in the shell-fish itself*. The finest pearls are always seen plunging into the body of the animal that inhabits the shell. I have remarked above that the pearl is a product of

---

* "It was probably from this spot," says Mr. Garner, "that Sir Richard Wynne obtained the pearl which he presented to the Queen of Charles II."

secretion ; it is a secretion of calcareous matter in a globular form under circumstances that are yet imperfectly known, though we can place the animal in a condition that will induce it to secrete pearls. For instance, if a specimen of *Unio margaritiferus* be taken, and one of the valves of its shell be pierced with a sharp instrument, so as to drill a hole *almost* through it, care being taken not to allow the instrument to penetrate completely through the shell, it will be found that the animal secretes a pearl upon that part of its shell.

Linnæus succeeded perfectly in causing the formation of pearls in the shell of this same freshwater mussel. He found that when grains of sand were placed between the shell and the body of the mollusc a pearl was produced which enveloped the grain of sand. This might have been expected, for sections of Oriental pearls often exhibit very fine concentric laminæ, surrounding a grain of sand, or some such extraneous matter.

We have only one or two more Bivalves to mention before closing this chapter.

Buffon speaks of a mussel found in the Mediterranean which the Sicilians and Italians turn to account for making gloves and stockings. It is a species of *Pinna*. This genus of mollusca belongs to the same group as the pearl oyster (*Avicula*); like other mussels, the *Pinna* secrete a long *byssus*, by which they hold to the rocks. The species vary much in dimensions according to their age, but often attain a considerable size, and secrete a byssus more than a foot long. The two valves of their shell are equal, and shaped somewhat like a lady's fan half open. Their byssus is not, like that of the common mussel, scanty and coarse, but long, fine, lustrous, and abundant. The animal lives generally half-buried in the sand, being anchored to an adjacent rock by its long byssus. The latter is not unlike silk, though its chemical nature does not appear to have been examined. It is employed in the manufactories throughout Italy. It appears that the Italians cannot dye this substance, and that, consequently, it can only be used in its natural brown colour. Reaumur called these mollusca the *silk-worms of the sea*. The inhabitants of Palermo have manufactured this byssus into various species of cloth, which are usually of a high price. It takes many individuals to furnish enough silky thread to manufacture a pair of stockings, and the thread is so fine, that a pair of stockings made of it can be easily contained in a snuff-box of ordinary size. The species generally sought for is *Pinna nobilis*, L. (Fig. 24, *P. marina* of others), which is taken off the coast of Sicily, at Toulon, etc., by means of a cramp, a species of iron fork, the prongs of which are perpendicular to the handle. It inhabits water from fifteen to thirty feet deep.

Pinna muricata has been called by the English "*the great silk mussel;*" and *P. flabellum* furnishes a similar silky byssus. These three species all inhabit the Mediterranean.

The genus *Pinna* is also remarkable by the fact that these mollusca, especially *P. nobilis*, produce pearls. These are generally small, and of an amber colour or reddish, sometimes grey or of a lead

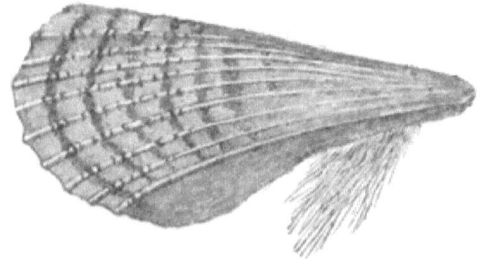

Fig. 24.—Pinna nobilis, L., showing its byssus, called by Reaumur the "Silkworm of the sea."

colour; others are black, and shaped like a pear. They are frequently large enough to be of considerable value.

* * * * *

The shells of these mollusca, which are not handsome enough to be employed in ornamental work, etc., can still be made useful in a variety of ways. They are composed of carbonate of lime, with a very little phosphate of lime and other salts, and organic matter. On soils which require lime, pulverized shells may be found of service, especially in vine countries, where lime in the soil has a marked influence upon the quality of the wine. By calcining them we obtain quicklime of a very pure description. By acting upon them with sulphuric acid, they are converted into gypsum or plaster of Paris (sulphate of lime), though this substance is too common in nature to induce us to prepare it in any quantity from shells. By dissolving shells in hydrochloric acid, after they have been calcined to destroy their organic matter, we can obtain chloride of calcium, a salt much used in chemical processes. By acting upon the lime produced from shells with chlorine, we can transform it into chloride of lime or bleaching powder, etc. All these products may be economically obtained from shells, such as the oyster shell, wherever they are abundant; and the compounds thus produced are purer than those obtained from chalk, or other varieties of carbonate of lime found in nature.

* * * * *

The beautiful molluscous animals included in the family of *Tunicata,* many of which resemble transparent bells of the most delicate organization, and some of which are phosphorescent at night, form valuable specimens for the aquarium. The *Bryozoa* are equally beautiful, but much smaller; and in many their beauties can only be appreciated under the microscope.

Chapter VIII.

Worms.

Curious observations upon Worms — Reproductive power of the Naïs—Sabularia—Terebella—Lumbricus—Planaria—Helminthes, or Entozoa—The common Earth-worm, Lumbricus terrestris—The Leech, Hirudo medicinalis—The Horse-leech, H. sanguisuga—Hirudiculture, or Leech breeding—Its cruelties—Extent to which it is carried on in France—Barometers of Leeches and Frogs—Worms for the Aquarium.

WORMS.

ONE of the most interesting classes of animals is certainly that of *Worms*. Who has not heard of the wonderful power of reproduction or regeneration of lost parts manifested by the *Naïs*, those curious little organisms which, in clusters of myriads upon myriads, form those large red patches on the muddy banks of the Thames or other rivers, and which vanish like magic when a stone or stick is thrown upon them? Cut off the head of one of these little fresh-water worms eight successive times, and you will find that it grows again seven times; the eighth decapitation has proved too much for the reproductive power of the *Naïs*, and this time the head has disappeared for ever! The number of times the head will be reproduced depends upon the vital powers of the individual submitted to experiment. Bonnet, in his "Observations sur les Vers d'eau douce," states that he cut a *Naïs* into twenty-six pieces, and each piece became a new worm. He produced thus twenty-six *Naïs*. He cut the head off the same *Naïs* twelve successive times, and twelve successive times the head was reproduced. M. Flourens, in his work " Sur la Longévité Humaine," etc., says, " There exists in the animal economy not only a *force of development* which brings each part up to the precise term assigned for it, but an individual *force of reproduction*, first brought to light by Trembley's experiments on polyps."

Look again at the marvellous manner in which the marine worms, *Sabularia* and *Terebella*, construct the tubes they inhabit, by means of the grains of sand and rock of the sea-shore, or at the curious *phosphorescent* faculty, or emission of light in the dark, possessed by many marine worms, and even by our common earth-worm (*Lumbricus*), at certain seasons of the year*; or still again, at the curious moveable organ of deglutition observed in certain voracious fresh-water *Planariæ*, which even after it has been torn away from the animal's body, continues to swallow down everything that is presented to its gluttonous orifice!

These worms may not appear to be directly useful to man, or to his commerce, save, perhaps, as articles sold for the *aquarium*, which has lately become so fashionable. But, on the other hand, what

* See my "Phosphorescence, or the Emission of Light by Minerals, Plants, and Animals." London, 1862.

a delightful and interesting source of study they afford us; and by such study are they not instrumental in enlightening our minds, in developing our pensive faculties, upon which the entire happiness of our race depends?

Greater marvels still await us in the numerous tribes of *Helminthes*, or intestinal worms. In these curious beings the organs of sense appear to be limited to that of feeling (or touch); in some digestive organs are altogether wanting, and their nutriment penetrates their tissues as it would those of a fungus or a conferva. No breathing apparatus is required here—how could it be otherwise with creatures who live constantly shut up in the tissues of other animals, often in cells or cavities which do not communicate with the external air? These curious animals are reproduced either by a sort of budding, by spontaneous division, or by eggs. When the two sexes exist, they are either found united on the same individual, or there exist distinct males and females. In these cases the young animal is developed from an egg; but between the egg period and that of the perfect animal, we observe, as in insects, mollusca, crustacea, and we may say, in fact, all other animals, a series of metamorphoses or transformations which, in the worms of which we speak, are exceedingly remarkable. Thus the embryo developed from the egg does not always grow up immediately into an animal similar to its parent. Often the young *helminthe* transforms itself into a species of *larva* capable of giving birth, without fecundation, to other *larvæ*, which are alone capable of becoming animals similar to the parent worm. But the most curious portion of their history is that these *larvæ* are generally found in the tissue of animals very different from the one in which the perfect worm exists, so that before one of them can complete its development, and become a perfect worm, it must be transported into another animal's body! Thus it is that *Cysticercus cellulosa*, Gm., which resembles a white cell or vescicle, and constitutes a peculiar disease with pigs, in whose muscular tissue it developes itself and multiplies with fearful rapidity, transforms itself into *Tænia*, or tapeworm, in the intestines of the human body; in fact, *Cysticercus* is the larvæ of *Tænia*.*

* But these details are foreign to my subject. I cannot, however, let pass this opportunity without noting down some recently discovered facts relating to this interesting class of animals. Among Helminthes, or *Entozoa*, as they are sometimes called, is a genus, *Filaria*, of which a species is often found in the heart of over-fed sheep, etc. It was formerly thought that these *Filaria* underwent no metamorphosis; but M. Joly has lately discovered a number of female nematoïd worms in the heart of a seal (*Phoca vitulina*); they belonged evidently to the genus *Filaria*: the individuals measured fifteen to twenty millimétres in length; the species appeared to be new, and was named *Filaria Cordis phocæ*. It is supposed that this worm is conveyed into the body of the seal by the fish which the latter feeds upon, and in whose bodies it exists in the larva state,

The only use that has yet been made of *Lumbricus terrestris*, or the common earth-worm, of which there are many varieties, is that of baiting the hooks and nets of fishermen. The large varieties that crawl upon the damp grass at night, living during the day in the earth, are captured in large quantities by poachers, etc., for baiting night-lines. In the same manner marine worms are used by the fishermen of seaport towns.

and is known at present as *Filaria piscium*. But this *F. piscium*, being always deprived of sexual organs, M. Joly looks upon it as the *larva* which, in the body of the seal, completes its development, and becomes *F. Cordis phocæ*.

Entozoa possess a wonderful tenacity of life. They have been known to revive after being placed for half an hour in boiling water. They have likewise been seen to survive the cold produced by ice; and they have been brought to life again after having lain in a dry state for six or seven years. They live in the most extraordinary places. In certain tropical climates there exists a species of rattlesnake, which, in Cumana, enters into the houses to catch mice. In the abdomen and in the large pulmonary cells of this reptile, a five-mouthed worm, *Pentastoma*, has been discovered. Another species of *Pentastoma* is found in the bladder of frogs. *Ascaris lumbrici*, a little spotted worm, the smallest of all species of *Ascaris*, has been discovered under the skin of our common earth-worm (*Lumbricus terrestris*), furnishing us with an example of a worm living upon a worm. *Leucophora nodulata* is a very minute worm, of a silvery or pearly aspect, living in the body of the small red worm, *Naïs littoralis*, of our river banks, and constitutes another example. These few notes will, I hope, show what peculiar interest attaches to this numerous and curiously diffused tribe of beings, and it is with much impatience that I await the forthcoming work of a truly able observer, Dr. T. Spencer Cobbold, upon this class of animals. Pouchet in his *Hétérogénie* energetically denies their wonderful migrations.

A worm which has attracted considerable attention lately, and by rearing of which large sums have been realized in France, is the leech (*Hirudo medicinalis*, L.)

Leeches are remarkable for their peculiar triangular mouth, which is provided with a lip, and their ten eyes. At the other extremity of their worm-shaped and extensible body is seen a kind of sucker, by which they adhere firmly to objects under water, whilst their head moves about in all directions. In many species two rows of pores are observed underneath the body; these pores are the orifices of so many small pouches, which constitute the animal's breathing apparatus.

The medicinal leech (*H. medicinalis*, L.), used for bleeding, is generally of a blackish colour, striped with yellow lines above and spotted yellow stripes beneath. It is found in all the still fresh-waters of Oriental Europe. The horse leech (*H. sanguisuga*, L.) is much larger, and of a greenish-black colour. It is common in our fresh stagnant waters.

The former species, *H. medicinalis*, has alone been submitted to special culture. In the countries where it is bred, it is reared in marshes specially adapted to that purpose; and until very recently its nourishment was derived from old worn-out horses, which, instead of being left to graze away in peace the last days of the weary life which they are forced to lead for man's comfort, were driven into the leech-ponds, to be fed upon by these noxious worms! Such, O readers! is the disgusting practice that has been followed in France for many years. This unwonted and unequalled cruelty constitutes a lasting disgrace to the Government which sanctions it. Very recently, however, the scientific men who form at the present time the most honourable portion of French society, and the most enlightened portion of its Senate, have begun to look with abhorrence at this frightful cruelty, and are endeavouring to prevent it. The *Société Protectrice des Animaux*, a most worthy institution, established in Paris, has awarded its silver medal to M. Borne, of Clairefontaine, and its bronze medal to Messrs. Harreaux, Sauvé, and Laigniez, for having abandoned this barbarous method of feeding leeches upon the blood of living horses, and for having constructed new marshes or leech-ponds, where the worms are fed with blood and other animal matters taken from the slaughter-houses.

For some years past, Messrs. Guénisseau and Fermond have been occupied with the culture of the leech; and M. Auguste Jourdier has recently published an interesting little work, entitled "Sur l'Hirudiculture,"* in which he treats of the rearing and artificial breeding of *H. medicinalis*.

* One vol. in 8vo, Paris, 1856.

To give some idea to what extent the breeding of this worm is practised in France, I may state here that a single *leech-swamp* in La Gironde yields, on an average, a return dividend of *fifteen per cent.!* Not long ago a similar marsh in the same district, and about 120 acres in dimension, sold for £10,000 sterling! I learn, moreover, from very reliable sources, that considerable fortunes have been realized in the neighbourhood of Bordeaux by breeding leeches.

But the day cannot be far off when all these leech-ponds will be dried up, and when the old barbarous practice of bleeding with leeches will be banished from a more enlightened medical generation. Then, indeed, will the useless cruelty of the leech-ponds vanish for ever, and no more old women or children shall be bled to death.

Some persons have attempted to convert the common leech into a barometer (Fig. 25). Among

FIG. 25.—Leech barometer.

other curious habits it has been observed, that on the approach of a tempest the animal ceases to be languid, moves about with a degree of activity "in proportion to the violence of the storm to come," and endeavours to escape by climbing up the sides of the glass jar in which it is confined. It is asserted that in this respect the leech is a dangerous rival to the little green frog, which is sold for a similar purpose on the Continent. A few of these frogs are placed at the bottom of a large glass vase containing moss, and half filled with water; a small wooden ladder reposes on the moss, and reaches to the top of the vase. When the weather is going to be calm, the frogs mount the ladder, and come and croak at the surface of the water; but when it is going to be stormy, they descend to the bottom, and bury themselves in the moss. But, for my own part, I do not place much reliance upon the indications of such-like barometers, and would advise my readers to adhere to that invented by Torricelli.

Since the aquarium has become a drawing-room ornament, or a living cabinet of natural history to the lovers of science, many species of worms, hitherto disregarded by the public at large, are fetching somewhat large sums in the market. Such, for instance, are certain *Serpula*, the beautiful organisms belonging to the genera *Sabella*, *Terebella*, *Spio*, *Sabularia*, etc., of which some of the

P

rarer species sell at very high prices. These worms, by their curious tubes or habitations, their gold-like branchiæ or gills, their curious habits, etc., are indeed objects most worthy of attention.

Chapter IX.

Polypes.

General remarks on Polypes — Their Organization and Polypidom — Naturalists who have written upon Polypes — Hydra fusca and H. viridis — Reproduction of Polypes — Polypes for the Aquarium — Corallium nobilis, and general observations on Coral — Its Polypidom — Practical details concerning Coral — Coralliculture — Coral Fishery — Uses of Coral — Isis hippuris, or Articulated Coral — Tubipora musica — The genus Madrepora — Reefs and Coral Islands — Formation of Reefs — Madrepora muricata — Its Chemical Composition — How it derives its Lime — Its uses.

POLYPES.

BETWEEN the class of Worms and that of Polypes there exists many groups of inferior animals which, hitherto, have not been employed by man; such, for instance, are the *Medusæ* (Sea-blubbers and Sea-nettles), and the different varieties of Star-fish (*Asteria, Ophiura,* etc.) Many of these are mentioned in my work on Phosphorescence, as most of them evince the faculty of becoming luminous in the dark. Some of these animals have been used as manure on the sea-coast, but with little or no effect. Among the Echinodermata (Star-fish, Ophiura, etc.) there is, however, an animal, *Holothuria priapus*, or sea-slug, which for years has been exported in large quantities from several of the Malay Islands to China, Cochin China, etc. Hundreds of junks or canoes are paddled along the shallow beaches on the coasts of the East India islands, and filled with these soft gelatinous beings. The *Holothuria* are purged of impurities by having quick lime thrown over them, dried in the sun, and packed in baskets, which sell at a high price among the Asiatics. Long before Polypes should likewise be placed the class of *Rotiferæ*, or wheel-animalcules; but, on account of their microscopic forms, the little I have to say upon them will be found in the chapter on *Infusoria*. The same remark will apply to some other microscopic beings.

Polypes comprise a numerous series of animals that have been classed in the genera: *Coralium, Isis, Madrepora, Caryophyllea, Oculium, Pocillopora, Astrea, Porita, Meandrina, Tubipora, Sertularia, Actinia, Hydra,* and a few others. They are wonderfully numerous. Nearly one-seventh part of the actual crust of our globe is composed of the remains of animals, and polypes contribute largely towards this fraction of our present world. Several species are valuable to us in different manners.

The body of a polype appears most simple in its organization; it consists of a little gelatinous sack or bag, the opening of which is surrounded by tentacles. Some species live separately, floating about singly in the water, or fixed one by one to the rocks: Others live in large companies, and secrete a curious habitation or basis, called a *polypidom*. They have been therefore divided into two groups, namely: Naked polypes, such as the *Sea Anemones* and the *Hydra* of our fresh-water ditches and ponds; and Coralligenous polypes—those which produce a polypidom—such as the *Coral*, the *Madrepora*, etc. The class was formerly much larger than it is now, and extended from Aristotle's polype—which is no other than the cuttle-fish, *Sepia octopoda* (*S. officinalis*)—to *Infusoria*, including animals which differ essentially in every respect. The habitation of Coralligenous polypes—the polypidom—was looked upon by the ancients as a growing stone or a stony plant (*Lithophyte*). The first observer who hinted at their *animal* nature appears to have been Imperati, and his observations, published in 1699, were confirmed by Peyssonel in 1727, and by Trembley about the year 1740, whilst engaged in his wonderful experiments upon *Hydra viridis* and *H. fusca* of our stagnant waters.

Ellis, Marsigli, Baster, Donati, Boccone, De Geer, Réaumur, De Jussieu, and Cavolini have added considerably to the interesting history of polypes. Linnæus called them animal plants (*Zoophytes*), and this celebrated naturalist classed the greater number of species, thus laying the groundwork for the later researches of Pallas, Bruguières, and Lamarck.

To Cavolini, Ehrenberg, and Savigny we owe much of our knowledge concerning the organization of corals; and for the description of the geographical distribution of islands, and other geological formations occasioned by these animalcules, we are indebted to the labours of R. and G. Forster, Chamisso (author of the "Marvellous History of Peter Schlemyll"), Peron, Quoy and Guemard, Captain Flinders, Lutke, Beechy, Darwin, D'Urville, and Lotin.

Alex. von Humboldt has sketched, in a charming manner, their influence upon the constitution of the earth's crust, in his "Views of Nature," vol. ii.

Hydra fusca, the olive-coloured polype of our ponds and ditches, may be taken as the type of this class of animals. This little being was first described by Trembley in 1744, but it had been previously discovered by Leuwenhoek in 1703. No attention was paid to it, however, till the publication of Trembley's paper, which produced great sensation, everyone's attention was drawn to the subject, and it became the principal topic of the day. It was given away in presents as an object of great rarity; specimens of it were sent from abroad by post, and even ambassadors made it a matter of engrossing interest in their relations to the foreign courts.

If a little duck-weed (*Lemna*) be put into a bottle of water with a wide orifice, and the bottle be placed upon a table, and allowed to remain perfectly still for some hours, the *Hydra* contained in the stagnant water will all come to that side of the bottle upon which the light falls, and will be seen floating about in that quarter of the flask, or adhering to that portion which is turned towards the window of the apartment. With a magnifying-glass it is easy to recognize *Hydra fusca*, which is brown or olive coloured, and *H. viridis*, which is green. Sometimes a reddish-brown variety (*H. rubra*) will be also seen. The little creatures appear like very small floating sacks, having four arms or tentacles spreading out from the orifice of the sack. If these animals be cut into several pieces with a scissors, each piece becomes a new hydra; if one of them be turned inside out like a glove, it lives so, the external part, which is now the interior, carries on the process of digestion as if it had always been inside.

Polypes are reproduced by "budding," by spontaneous division, or by eggs. In the first process one or more buds form around the mouth (orifice of the sack), or on some other part of the animal's body. This bud, which at first appears as a little globule, gradually developes itself into a complete polype, and drops off. This process of reproduction is extremely rapid; a single day often suffices for several successive generations to make their appearance. Thus, a child polype born by budding at six o'clock in the morning, will, in many cases, be a grandfather by six in the afternoon. But this rapid succession of births is only observed in all its grandeur under the Tropics. It has been remarked, also, that the larger species of polypes produce fewer young.

The *Hydra* that live in the ditches and stagnant ponds around London, Paris, etc., die in the winter; but before this their body is replete with eggs or buds, which are dispersed in the water in the form of minute granular bodies, to become new polypes the ensuing spring. These fresh-water polypes are interesting objects of study for the fresh-water aquarium, and as they are of a certain size, they can be easily observed by means of a common lens or magnifying-glass. It is curious to see them seize in their tentacles small worms, insects, etc., and carry them into their semi-transparent gelatinous body.

The same may be said of the *Flustra*, which belong to the higher class of *Bryozoa*, and form interesting specimens for the salt-water aquarium. Many varieties of them are found on the sea-weeds, shells, rocks, etc., which they cover with a minute network of cells. Each cell contains a polype-like animal, and there are in some species many hundred cells in one square inch of this network. Again, the *Sertularia* and the beautiful *Campanularia*, or bell-shaped polypes, are sought for to decorate the aquarium; whilst *Sea Anemones*, on account of the

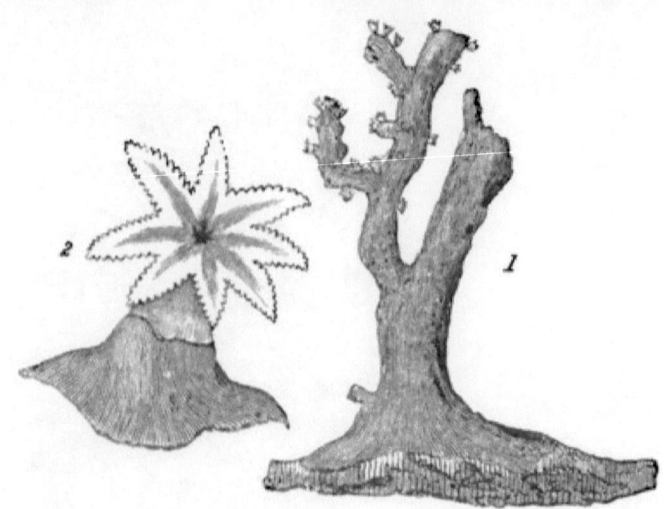

Fig. 26.
1. Corallium nobilis (Red coral).
2. Polype magnified.

comparative ease with which they are reared, form frequent and interesting objects of study in the same miniature ocean.

Polypes have numerous enemies in the shape of worms, crustacea, fish, water insects, etc. They also devour each other when opportunity offers, but it has been observed that polypes of the same species cannot digest each other.

They appear to live principally upon animal substances, such as small worms, infusoria, and the like, with which the waters they inhabit generally abound. Certain sea anemones have been seen to devour small fish; in the aquarium they are fed with small pieces of raw beef.

Some polypes remain for ever attached to their cells, and cannot be drawn from their polypidom without being killed. Others appear capable of leaving their habitation, to wander about and construct another polypidom at some distance from the old one; but this fact has not been sufficiently proved.

The most important polype, in a commercial point of view, is the Coral (*Corallium nobilis*, L. Fig. 26); the bright red substance of its polypidom has rendered it valuable as an article of trade. After pearls, coral is considered the most precious production of the ocean, and on the coasts of the Mediterranean it has for ages been the object of an extensive traffic. In nature its stem, or the axis of its polypidom, is calcareous, solid, and striated; it is covered by a gelatinous porous envelope, in which the coral polypes are seen implanted.

Donati has thrown much light upon the organization of the coral stem, and the anatomy of the gelatinous tunic which covers it, and places each of its polypes, as it were, in connection one with the other. It will be sufficient here to state that the coral polypes produce the calcareous portion of their polypidom, and also secrete this gelatinous covering, which is of a very complicated nature. The latter, when the coral is freshly taken from the water, is easily peeled off; but if allowed to dry on the stem, it becomes very difficult to detach it. This cortex, or covering, presents numerous tubercles or little eminences, each of which contains in its cavity a white, soft, transparent polype, having eight tentacles. As soon as the coral is withdrawn from the water, each polype immediately contracts itself, and withdraws into its cavity.

The external portion of the solid coral stem is generally much less compact than the interior. When calcined, it loses its organic matter and its colour, and is then seen to be composed of concentric layers. Silliman, jun., has analyzed this substance; he finds that it is composed of carbonate of lime, containing three to five per cent. of organic matter, and very small quantities of silica, fluoride of calcium, fluoride of magnesium, phosphate of lime, alumina, and oxide of iron. The red colour I believe to be entirely organic, though nothing is yet known concerning it; and though coral is generally of a fine red colour, it is sometimes found of a rose tint, or even quite yellow. There is also a black variety, which is very rare. Its gelatinous tunic also varies in colour.

The calcareous stem of these animals is formed like the shell of the oyster and other mollusca, *i.e.*, by the secretion of a liquid containing a large amount of lime, and which appears to be produced by certain glands situated at the basis of the polype's tentacles.

In the Red Sea and the Mediterranean, coral is seen adhering to the rocks in all directions. The greatest height that a stem of coral, with its branches, will attain in the Mediterranean is about a foot and a half, its greatest diameter being about eight lines.

At each extremity of the coast of Algiers very fine coral is found. The annual production by coral fisheries in these parts is estimated at about £100,000 sterling. But the French are complaining, at the present moment, of the negligent manner in which their Mediterranean coral production is carried on. It should yield, according to competent authorities, *a nett profit of* £250,000 *sterling per annum.**

Spallanzani's observations have taught us that coral grows very rapidly, and is quickly reproduced; so that in a few years' time a locality which has been deprived of its coral by repeated fisheries is again repeopled with this lucrative polype.

It has also been remarked that a branch of coral, detached from the stem and thrown into the sea, soon fixes itself to the rocks, and grows into a fine specimen; and it has not unfrequently been noticed that different objects which have been thrown into the sea near any clusters of coral, are sure to be covered with these polypes in the course of a few months.

These important facts seem to indicate the possibility of transporting or transplanting the coral by shoots, as we do with some of our rarer vegetable productions. They teach us, also, that the coral fishers ought to be compelled by law to throw back into the sea the younger branches of whatever coral they take away; for these young shoots are nearly valueless to them, and would serve to replenish in a short time places exhausted of their coral by constant fishing.

Like other polypes, the coral polype is repro-

* Compare the "Bulletin de la Société d'Acclimatation," Paris, 1856.

duced by eggs, by buds, and by self-division. It multiplies rapidly, and its stem will go on ramifying, like the stem of a tree, for an indefinite period of time.

All these data should be borne in mind by those who would undertake to cultivate coral, a branch of industry which has lately been seriously thought of, and to which the French have already given the name of *Coralliculture*. And if it be impossible to grow coral upon our English coasts, there are spread over the globe hundreds of English possessions where *Coralliculture* might become an unexpected source of wealth.

For ages past coral has been the object of an extensive and valuable industry; it constitutes an important feature in the commerce of Marseilles, Genoa, Catalogna, Corsica, Sicily, and other Mediterranean islands. The coasts of Sicily, the Adriatic, and the coast of Tunis, are classed among the places where the most active operations of this kind are carried on. Regular coral fisheries are established in the Straits of Messina, on the shores of Majorca and Ivica, the coasts of Provence, of Algiers, etc. Abundant supplies are obtained from the Red Sea, the Persian Gulf, the coast of Sumatra, and other localities.

Sicilian coral is much prized, and has been known to value as much as ten guineas per ounce.

Q

The price, however, is exceedingly variable, according to quality, other portions of the same mass selling for less than a shilling a pound.

Coral fishery takes place during the three hottest months of the year; the only instrument that the fishers employ is the *salabre*, a kind of dredge, consisting of two strong sticks crossed one over the other. To the centre of the cross is a long rope, and underneath it a bullet or stone. At the four extremities of the sticks, which are covered with tow (hemp), is a net shaped like a purse (Fig. 27).

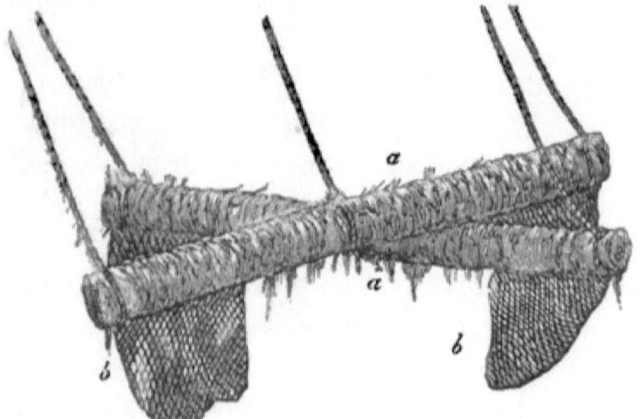

FIG. 27.—Coral Net.
a a. Beams of wood, 15 feet long, covered with tow.   *b b.* Coarse nets.

This instrument is dragged over the rocks from which the coral springs, and the latter broken off by the dredge, its branches become entangled in the tow, and are secured by the net. But by this clumsy apparatus, as our readers will easily conceive, a great quantity of coral would be lost, were it not sought for immediately afterwards by **divers**, which is generally the case. This **fishing or** dredging generally takes **place** at a depth **varying** from sixty **to eighty feet, but coral is sometimes** dredged for **and taken at upwards of one** hundred feet below the **surface of the sea.**

In Europe, particularly at Marseilles, **coral is manufactured into** a great variety of ornaments; it **is also** largely dealt with in the East, in India and **Africa,** where it is employed to ornament weapons, for jewels, chaplets, etc. When the Arabs bury any of their relatives, they always place in the dead person's hand a chaplet **of coral.**

In Europe **coral used also to be** employed in medicine, but **it has been found that** a little limestone serves **the same purpose. It is** extensively **used** for jewellery, **and is also made** into toothpowder.

In 1852, the quantity of red coral imported from **Italy to** Liverpool amounted to 120 lbs.; in 1854, **146 lbs.** arrived.

There exist four species **of coral-like** animals belonging to the genus *Isis* (which has been separated from that of *Corallium*), **one of** which, *Isis hippuris*, know as *Articulated coral*, is abundant in many seas. Its polypidom is composed of calcareous joints united to and alternating with horny ones, which gives to the species in question an aspect similar to that of the plants called *Equisetum* (horse-tail). *Isis hippuris* is sought for and prized as a curiosity, though the species is not rare.

The polypes of the genus *Tupipora* are extremely remarkable, and much prized as curiosities. Their polypidom is composed of a series of bright red calcareous tubes or prismatic cylinders. They form large round tufts, and often considerable masses in the warmer seas. Péron found that the polypes that inhabit these tubes have green tentacles, so that large agglomerations of these species appear like tufts of grass or green fields in the ocean.

The species *Tupipora musica* is the most common; its polypidom is of a fine red colour; it has been termed *T. musica* because the cylinders of this polypidom call to mind the tubes of an organ. It is found abundantly in the Indian Ocean and American seas. Formerly it was employed as a medicine, but now is only sold as a cabinet ornament or a curiosity.

It would be interesting to cultivate the latter two, and several other allied species, in a warm salt-water aquarium. Such an aquarium might be easily established in the warm greenhouse of Kew and other botanic gardens, and it should contain some of the rarer marine *Algæ* along with these magnificent polypes.

It is to the genus *Madrepora* that most of the so-called "coral-reefs" are owed. Every one knows how dangerous these reefs prove to navigators, and what an extensive part they play in the constitution of the earth's crust. Their colours are almost invariably white or yellowish-white; but there are some which are completely yellow, red, or brown. These *Madrepora* are extremely common in nature, and abound near the islands of the South Sea, of the Indian Sea, and especially near the Antilles. Captain Cook tells us "that he could not sail through certain straits which he had passed with ease a few years previously, on account of the prodigious and rapid multiplication of these coral-reefs." There is a barrier reef of madrepores that runs along the whole of the eastern coast of Australia. Captain Flinders endeavoured for fourteen days to pass through it, and he found that he had sailed more than five hundred miles before he accomplished his purpose. Throughout the whole range of Polynesian and Australian islands, there is hardly a league of sea unoccupied by a "coral-reef" or a "coral-island."

These reefs develop themselves in proximity to the shores of continents and islands, or upon the summits of submarine volcanic rocks. The latter circumstance explains the frequency of their crater-like forms (Fig. 28). Dalrymple says he has seen

Fig. 28.—Circular Coral Island, recently formed in the Pacific Ocean, principally composed of the species *Madrepora muricata*, and shutting in a portion of the ocean as a lake.

madrepore banks in all their stages—some in deep water, others with a few portions above the surface; some just formed into islands without the slightest vestige of vegetation; others with a few weeds on their highest point; and, lastly, such as are covered with trees of many years' growth, "with a bottomless sea at a pistol-shot distance."

As soon as the edge of a reef is high enough to lay hold of the floating sea-weed, to retain the seeds of plants brought by the winds and currents, or for a bird to perch upon, the "coral-island" may be said to commence its existence. The excreta of birds, wrecks of all kinds, feathers, cocoa-nuts floating with the young plant out of the shell, various grains, and sea-weeds, are the first elements of the new island.

With islands thus formed, and others in the several stages of their formation, Torres Strait is nearly choked up. The time will come—it may be ten thousand or ten million years, but come it must—when New Holland and New Guinea, and all the little groups of islets and reefs to the north and north-west of them, will either be united in one great continent, or be separated only by deep channels, in which the strength or velocity of the currents may perhaps obstruct the silent and unobserved agency of these insignificant, but most efficacious labourers.

Fig. 29.—Fragment of Madrepora muricata.

Madrepora muricata, L. (Fig. 29), is the species which contributes most largely to the formation of reefs; it is often sold for ornaments, particularly in

France, where it is called *Corne de Dame*, or *Char de Neptune*. There are some splendid specimens of this and its allied species in the British Museum. Immense masses of its beautiful and wonderful structure are employed to manufacture lime for building and manure. The inhabitants of the Polynesian and Australian islands burn it to produce the lime with which they chew their betel, and scour the *Holothuria* which they collect for the Chinese, etc., as we have already seen. The lime thus produced is very much superior to any that can be obtained from lime stone, however pure. When employed as manure, it would be better to crush it without burning it, as it would thus retain its animal matter; but some varieties are so hard, that the crushing can only be effected with very powerful machines. *Madrepora* and other closely-allied polypes—such as *Porita, Astræa, Meandrina, Caryophyllea* (Fig. 30)—contain from 90 to 95 per cent. of carbonate of lime, with a little carbonate of magnesia; they also contain a very small quantity of fluoride of calcium and phosphate of lime, which latter, small as the quantity is, renders them still more valuable for agricultural purposes.

An analysis which I made of *Madrepora muricata*, in 1859, gave me 5 per cent. of organic matter, 0·4 of silica, 92·27 of carbonate of lime, 0·69 of carbonate of magnesia, 0·65 of phosphate of lime, oxides of iron and alumina, 0·99 of sulphate of lime, and traces of fluoride of calcium.

All these salts are extracted, by the polypidom-making polypes, from the water of the sea. If we

Fig. 30.—Caryophyllea fastigiata.

analyse the water of the ocean near "coral-reefs," we find a considerable deficiency of lime. Thus, Dr. Forchhammer, in an interesting paper, has lately shown that where madrepore polypes abound, the salts furnished by the sea only contain 2 per cent. of lime. But, on the other hand, these polypes can never extract *the whole* of the lime from the sea-water, as this author and others appear to think, for Nature has established here one of her beautiful *rotations:* as the little polypes extract lime from the water to form the *new* portions of their poly-pidom, the water, by means of the carbonic acid it contains, and with which it is supplied in great measure by the polypes themselves, dissolves the more ancient portions of their calcareous structure, thus keeping a constant supply of carbonate of lime at their disposal in the water.

In the South Sea Islands, the madrepore structures are occasionally employed as building stone; they are known as *coral-rock*.

Madrepora was formerly imported into this country for medicinal purposes, under the name of *white coral*. It is capable of receiving very fine polish, and can then be made, as coral, into ornaments of every description.*

---

* For many extremely interesting and novel details concerning fresh-water polypes, bryozoa and infusoria, see Henry J. Slack's ingenious little work entitled "Marvels of Pond Life."

Chapter X.

Infusoria and other Animalculæ.

Microscopic Animals useful to Man—Universal distribution of Infusoria—Dry Fogs—Authors who have studied Infusoria—Philosophical considerations concerning them—The Monads, Rotifera, Vibrio—Rhizopoda—Monas crepusculum, the most minute of living beings—Deposit in which the Transatlantic Cable lies—Transition of Colour in Lakes—Fossil Infusoria—"Mountain Meal"—Its Chemical Composition—Enormous quantities of it consumed as Food—Geographical distribution of Infusorial deposits—The Town of Richmond, in Virginia—Berlin—The Polishing Schist of Bilin, in Prussia—1,750,000,000 beings to the square inch—Tripoli, its uses and composition—Geographical and Geological distribution of Infusoria, Foraminifera, and Diatomaceæ—Soluble Glass obtained from Infusorial Deposits—Uses of Soluble Glass—Other applications of Infusorial Earth—Chalk, its uses and geological origin—The Nummulite Limestone—Paris mostly built of Animalculæ—Other details—Time.

INFUSORIA AND OTHER ANIMALCULÆ.

WE pass on now to examine another extensive group of animals, still more wonderful, and perhaps more interesting, than any which precede. Here, under the highest magnifying power of the microscope, we find animals useful to man—here, amidst the millions of invisible atoms which nature has so abundantly scattered over the globe, we find delicate and wonderful organisms, supplying us with food, with pure water, with glass, with colours, and last, not least, with an inexhaustible field of scientific inquiry. Look where we will, we find them everywhere—in our bodies, in our aliments, in our drinks, in our preserves, in the water in which we bathe, on our walls, on our glazed paper, on our visiting cards, on our flowers, in the soil of our gardens, in the woods and forests, in our meadows and their trenches, in our ditches, ponds, lakes, rivers, seas, and oceans, in the oldest sedimentary strata of the earth, in the most recent strata, on the mountain tops, in the snow and in the ice, and sometimes in the air we breathe.

Ehrenberg found a few species of *Infusoria* in the subterranean water of mines; he met with several in some silver mines in Russia, at the depth of fifty-six fathoms below the surface; but he never detected them in atmospheric water, such as dew-drops.* The same author discovered that the yellow dry fog which has been observed from time to time advancing from the Cape Verd Islands towards the east, covering parts of North Africa, Italy, and Central Europe, is composed of hosts of silicious animalculæ, carried away by the trade-winds. This peculiar meteor has been often attributed to the tails of comets which have passed near the earth's orbit.† Similar animalculæ have been found in fixed or floating icebergs at 12° lat. from the North Pole, while numerous forms of the same group are seen in hot mineral springs.

The invention of the microscope by Hans Jan-

---

* This observation, made many years ago, agrees admirably with the results of numerous researches lately made by Pouchet of Rouen, who discovered no infusoria in snow that had recently fallen, nor in the atmosphere. It has been held that the air abounds with eggs of infusoria and seeds of microscopic plants; but Pouchet denies this, upon the strength of many experiments made in various parts of Europe.

† See Humboldt's "Views of Nature," tome ii.; also Kaemtz's "Meteorology," and my work on "Phosphorescence," pp. 55-57, regarding the nature of dry fogs.

sen and his son Zacharias Jansen of Middleburg, revealed to us the existence of myriads of living creatures, of whose presence in nature we had not before the slightest suspicion; and observation has disclosed a number of organic creations comparable only to that of the stars revealed by the telescope. When Linnæus arranged all the organized beings known to him in his " Systema Naturæ," the structure of infusoria and other animalculæ was not sufficiently known to enable him to distribute them properly. He therefore placed them at the end of his last class, *Vermes*, in a genus which he denominated *Chaos*.

Othon Frederic Müller first distinguished them as a distinct order, and finding they were so quickly produced in infusions of vegetable substances, called them *Infusoria*. Müller's work was published in 1773-4. He described many species. But Needham had already published (1745) his " New Microscopical Discoveries."

These minute organisms have also been investigated by Leuwenhoek, Lamarck, Cuvier, Bory de St. Vincent, Hill, Hooke, Adams, Baker, Spallanzani, Ehrenberg, Mantell, Pritchard, Morren, Pouchet, etc.

Ehrenberg studied their internal structure by feeding them on colouring matters, such as indigo, and carmine.

If a few flower stalks or a handful of green leaves be placed in a glass of water, and allowed to remain there from two to four days exposed to the air and to the light, at the end of that time the water will have assumed a green or brownish-green colour, and on being submitted to examination under the microscope, will be found to swarm with many descriptions of infusoria. How they come there is still a subject of discussion among many of the first men of the day. Some say their eggs or "buds" are constantly present in the air, driven about everywhere by the wind, and develop themselves whenever they happen to fall upon an appropriate medium, such as putrefying vegetable substance, etc. Others say that no such eggs are present in the air, but that they form spontaneously in water containing vegetable matter, as the eggs of other animals form in the womb.*

Lamarck, Oken, Geoffroy St. Hilaire, Bory de St. Vincent, Darwin, and other distinguished naturalists, look upon certain infusoria (*Monades*) as the fundamental organic substance from which all higher organisms have been progressively developed. Nature created *Monades*, the most simple form of infusoria, from the gradual perfection of which, through myriads of centuries and amidst all kinds of physical changes, all the higher classes of animals

* Pouchet "Sur l'Heterogenie," Paris, 1859, 1 vol. in 8vo.

have been produced.* I myself have shown recently how mineral matter can be converted by chemical means into organic matter, and how this organic matter, in the origin, must have been converted into organized cells.†

"In vain," says Bory de St. Vincent, and his words coincide remarkably with our modern researches, "in vain has matter been considered as eminently *brute* [without life]. Many observations prove that if it is not all active by its very nature, a part of it is essentially so, and the presence of this, operating according to certain laws, is able to produce life in an agglomeration of the molecules; and since these laws will always be imperfectly known, it will at least be rash to maintain that an infinite intelligence did not impose them, since they are manifested by their results."

But we must quit these philosophical considerations, as our work is purely of a practical nature. Let us see then, first, what *Infusoria* are, and how they are useful to man.

The most simple and commonest form of infusorial life is the *Monad*. This animalcule, of which there are several kinds, consists of a fine pellucid membrane; it forms a very minute sphere

* Darwin "On the Origin of Species by Natural Selection," London, 1860.
† Phipson "Protoctista," etc., in the "Journ. de Medicine," Bruxelles, Dec. 1861.

R

or cell, having a few green or coloured spots in its interior. These curious beings are very small; I never measured any, but I find they require to be magnified at least 640 times to be seen at all distinctly. Some authors say they vary from 1-24,000th to 1-500th of an inch in size, according to the species. In the opinion of Humboldt, the true monad never exceeds 1-3000th of a line in diameter. He alludes probably to *Monas crepusculum*, the smallest species. One single drop of water may contain about 500,000,000 monades, a greater number than our earth contains of human inhabitants.*

They effect their locomotion by means of *cilia*, fine hair-like processes which cover the whole surface of the animalcule's body, and which are constantly vibrating, like those which are found on several membranes of our own bodies. Such is the

* Even in Leuwenhoek's time the excessive number of animalculæ in some waters was noticed with surprise; but in his day the microscopes were exceedingly defective. The eminent naturalist Swammerdam, who published the results of his dissections in 1660, had to work with very imperfect glasses. Leuwenhoek, who made known his curious and novel discoveries about 1677 (some years before and after), laboured under the same disadvantages. He actually ground his own lenses, in which art he excelled the best opticians of the day. Most of his papers have been published in the English "Philosophical Transactions." In a paper of his published in the "Philosophical Transactions" for 1677, we are struck by the ingenious method he employed to calculate the number of animalculæ present in a drop of water.

type of Infusoria in general; but there are other more highly-organized forms in this vast family, which recall sometimes the bell-shaped polypes, or other animals of still more complicated structure. The *Rotifera*, or wheel-animalcules, which were until lately classed with Infusoria, have been gradually elevated to the class of *Worms*, and are now placed by some zoologists near the tribe of mites (*Acarus*). They belong, therefore, to the highest of inferior animals, namely, to the class of Spiders. The *Vibrio tritici*, an eel-like animalcule, which causes the "earcockle," or the blight, in wheat, has been taken from the class of Infusoria, and placed in that of *Helminthes* or *Entozoa* (worms).

Some infusorial animalculæ secrete themselves a covering of hard flint (*silica*), resembling in this respect the plants which belong to the family of *Equisetaceæ* and the *Grasses*, the epidermis of whose stems contains sometimes as much as 90 per cent. of silica.

The covering or outer tunic of Infusoria is, then, of two kinds: the one soft and apparently membranous, yielding to the slightest pressure; the other rigid and hard, having the appearance of a shell, though, from its flexibility and transparent nature, it is more like horn. The microscopic beings belonging to the class of *Rhizopoda*—a class higher than Infusoria—present also the latter peculiarity. This hard covering consists sometimes of silica, and sometimes of carbonate of lime. To it we owe the preservation of the forms of *Infusoria* and *Foraminifera* (Rhizopoda), which have lain for centuries upon centuries in a fossil state in the strata of the earth. It has been calculated that eight million individuals of *Monas crepusculum* can exist within the space that would be occupied by a single grain of mustard-seed, the diameter of which does not exceed the one-tenth of an inch.

Yet these myriads of little beings termed *Infusoria* have an important part to play in nature; they help to keep the water they inhabit in a pure state. They devour animal and vegetable matter which otherwise would ferment, decompose, and render the water putrid and unwholesome for the use of superior animals.

The flint-shelled infusoria, together with numerous groups of lower beings (*Diatomaceæ, Desmidiæ*, etc.) and the *Foraminifera*, form after death considerable deposits at the bottom of the ocean—deposits which increase every day. In such a material lies the transatlantic telegraph cable, and by the progressive accumulation of these minute organisms deprived of life, and the gradual precipitation of carbonate of lime, clay, etc., from the water of the sea, the now soft muddy deposit thus formed will, in course of time, become a hard rock.

It is our hope to have a telegraphic cable, uniting us with the continent of America, imbedded one day in such a rock, where it would lie securely for ages. (See Fig. 37.)

The rapid and mysterious transition of colour which is observable in lakes, and which has often created alarm in the minds of the superstitious, has been attributed* to Infusoria. A lake of clear transparent water will assume, for instance, a green colour in the course of the day; it will become turbid or mud-coloured about noon, when the sun brings the Infusoria to the surface, rapidly develops them, and where they die by millions before night. Microscopic vegetables (*Algæ*, etc.) may produce similar effects. Similar phenomena are observed in salt water; hence, probably, the Red Sea and Yellow Sea derived their names. Certain *Astaria* and *Euglena ruber* give to water a blood-red colour. The same happens when microscopic *Algæ*, of a red tint, found at certain seasons in the Red Sea, are present. *Euglena viridis, Cryptomonas glauca, Monas bicolor*, and other Infusoria, colour water intensely green. A blue colour will be observed when considerable quantities of *Stentor ceruleus* are present, and yellow with *Astaria flavescens* and *Stentor aureus*, etc. Of these the green and red tints are the most frequently seen in nature.

* By Pritchard and others.

Again, many *Infusoria* and *Rhizopoda* play an important part in the phosphorescence of the sea. The luminosity of the waves is entirely due to them.

Ehrenberg has detected an immense number of fossil Infusoria (Fig. 31). At first they were found principally in certain siliceous deposits near Berlin, but they were afterwards recognized in all parts of the globe. Most of the species are so admirably preserved, on account of their siliceous and imperishable envelope, that they can be, at the present day, minutely investigated and classed.

These shell-like teguments of beings, invisible to the naked eye, are found in large masses, covering many miles of the earth's surface.

They constitute masses of a delicate white powder, known as *Mountain meal* (*Berg-mehl*, Germ.; *Farine de montagne*, French).

In Swedish Lapland, under a bed of decayed moss, forty miles from Degesfors, in Umea Lapmark, is found an immense stratum of this substance. Chemical analysis shows it to be composed of 22 per cent. of organic matter, 72 per cent. of silica, 6 of alumina, and 0·15 of oxide of iron.*

In times of scarcity, this "mountain meal" is mixed with flour, and manufactured into bread for the poor. These fossil Infusoria do not constitute

* This analysis was executed by Dr. Trail.

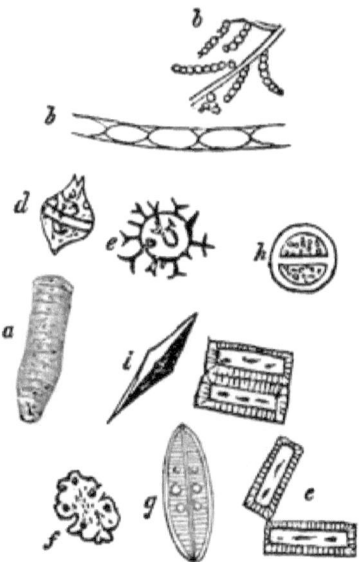

FIG. 31.—Fossil Infusoria, as seen (highly magnified) in the Berg-mehl.

a. Gomphonema.
b.b. Gallionella.
c. Bacillaria.
d. Peridinum.
e. Xanthidium.
f. Euastrum.
g. Pinnularia.
h. Pixidula.
i. Navicula.

of themselves an aliment of sufficient nutriment to sustain life; but in China, where "mountain meal" abounds in some districts, the poorer classes can, by its means, subsist twice as long upon the same supply of provisions as they could do were they not to make use of it.

This farinaceous substance consists principally of the remains of infusoria and microscopic vegetables. Under the microscope we recognize in it *Navicula viridis, Gallionella sulcata, Gomphonema gemmatum*, and several other species.

Berzélius and Rétzius affirm that, at the extremity of Sweden, the peasants are in the habit of eating this infusorial earth to such an extent, that every year *many hundred cart-loads* are extracted by them from the strata in which it is found. Some eat it from habit or taste, as we smoke tobacco; others from pure necessity.* Certain deposits of this kind serve for other purposes, as we shall see presently.

In America, deposits of infusorial earth have been discovered at West Point; then at Connecticut, Rhode Island, Massachusetts, and Maine, in which provinces no less than thirteen localities have been found where this "mountain meal" exists. Some of them have as much as fifteen feet in

* Compare with this Humboldt's "Views of Nature," vol. i., on the earth eaten by the Otomacs, etc.

thickness. There are seven or eight similar deposits in Mexico. All these deposits contain a certain amount of vegetable remains. Indeed, a similar kind of earth, composed almost entirely of microscopic plants (?) (*Diatomaceæ*), underlies the town of Richmond, in Virginia, North America; and the layer upon which this town is built has a thickness of no less than twenty feet.

The guano deposits of Ichaboe, and indeed all other beds of this substance, abound in remains of animalculæ and inferior algæ.

In some mud brought from the Levant, in 1844, hundreds of siliceous shells of Infusoria, Diatomaceæ, etc., were discovered; and some earth recently found near Newcastle, in England, was found to be almost entirely composed of fossil *Infusoria* and *Bacillaria* (minute organisms that some naturalists consider as plants, others as animals).

Moreover, some specimens of siliceous rock, from the Isle of France, were found by Ehrenberg to consist principally of fossil *Infusoria*, identical with certain living species.

In some of the plains of Eastern Germany such infusorial deposits are both common and extensive. The town of Berlin is built upon one of them, which measures about twenty-five yards in thickness. But it is a curious fact that the deposit which underlies the town of Berlin is composed of *Infusoria* and *Diatomaceæ* which are still living, and propagate daily with astonishing rapidity. Their existence is doubtless maintained by the waters of the river Spree, situated on a higher level, which filter into the deposit. It is feared that a period will arrive when a part, at least, of the town will fall in, on account of the rapid development of these microscopic creatures, more especially the *Gallionella*, which, according to Ehrenberg, form, in the space of four days, no less than two cubic feet of new movable earth.

The "polishing slate" of Bilin, in Prussia, which is used for polishing metals, glass, marbles, etc., forms a series of strata fourteen feet thick. It is entirely composed of the siliceous shells of *Infusoria* and *Diatomaceæ*, among which the most common appear to be *Gallionella distans* and *G. ferruginea*. One cubic inch of this polishing earth has been shown, by accurate measurement and calculation, to contain 41,000,000 individuals of *G. distans*, and 1,750,000,000 individuals of *G. ferruginea* (Figs. 32 and 33). In the present state of physiological science it is impossible to say whether these wonderful organisms are plants or animals. They furnish us with an admirable polishing material, for which it would be difficult to find a substitute.*

* These and other fossil animalculæ may be purchased in London, from the different dealers in minerals, etc. Their structure can only be discerned under a good microscope.

Under the name of *Tripoli* are included several of these siliceous infusorial earths, extensively em-

Fig. 32.—Gallionella ferruginea.
1. Magnified 300 times.   2. Magnified 2000 times.

Fig. 33.—Gallionella distans.

ployed for polishing metallic surfaces, etc. They derive their name from Tripoli, in Barbary, whence the substance was originally procured.*

Is it not an interesting fact that the remains of creatures individually invisible to the naked eye, should, in course of time, form rocks and strata destined to figure among the economical applications of the human race?

Since 1836, Ehrenberg has observed that the organic forces are still so active in the mud of ports and rivers, that at Swienemünde, in the Baltic, for instance, where more than two and a half millions of cubic feet of mud were recently

---

* Some kinds of *Tripoli* are entirely mineral, but these are generally known as *Emery*.

removed in one year, one-third of that entire mass consisted of microscopic animals. The moors of Limburg present accumulations of fossil Infusoria twenty-eight feet in thickness. In the peaty layer of Berlin, funnel-shaped deposits of Infusoria reach, in some places, to the depth of sixty feet. There is no doubt that they are still alive, and capable of increase. Spontaneous motion may often be observed in specimens taken from the greatest depth, though less frequently than in those taken from the surface.

The antiquarian, in bringing the microscope to bear in his researches, and by the discovery of these siliceous shells of Infusoria in various ancient articles of pottery, and the remains of similar species in the clay of the vicinity in which they occur, has proved that these vases were made upon the spot, and not imported from the higher civilized nations of that day, as had been previously supposed. In like manner thieves have been tracked and robberies discovered by means of the fossil Infusoria adhering to the boots of the suspected persons, though the latter had travelled many miles from the spot where the act was committed.

These fossil Infusoria and Diatomaceæ are found to belong both to marine and fresh-water species; many of them are in every respect identical with species still living. Their geographical distribution, and that of the equally microscopic but much larger *Foraminifera*, is remarkable by its extent.

"Not only in the polar regions," says Ehrenberg, "is there an uninterrupted development of active microscope life, where larger animals can no longer exist, but we find that the microscopic animals collected in the Antarctic expedition of Captain James Ross exhibit a remarkable abundance of unknown and often most beautiful forms. Even in the residuum obtained from the melting ice swimming about in round fragments in latitude 70° 10′, there were found upwards of fifty species of siliceous-shelled *Polygastria* and *Coscinodiscæ*, with their green ovaries, and therefore living, and able to resist the extreme severity of the cold. In the Gulf of Erebus, sixty-eight siliceous-shelled *Polygastria* and *Phytolitharia*, and only one species of a calcareous-shelled *Polythalamia* (Foraminifera), were brought up by a lead sunk to a depth of from 1242 to 1620 feet."

Dr. J. Hooker found siliceous *Diatomaceæ** in countless numbers between the parallels of 60° and 80° south, where they gave a colour to the sea, and also to the icebergs floating in it. The death of these organisms in the South Arctic Ocean is producing

---

* The *Diatomaceæ* are vegetables for some authors, animals for others. See on this subject my paper entitled *Protoctista*, cited on p. 241 of the present work.

a submarine deposit, consisting entirely of the siliceous particles of which the skeletons of these inferior beings are composed. This deposit is seen on the shore of Victoria Land, and at the base of the volcanic mountain Erebus.

Samples of water taken up by Schager to the south of the Cape of Good Hope in 57° lat., and again under the tropics in the Atlantic, show that the ocean, in its ordinary condition, and without any apparent discoloration, contains numerous microscopic living organisms. Ehrenberg has shown that the infusorial beings now living flourish at heights of 10,000 feet on land, far above the snow level, and at depths of 10,000, 12,000, and 16,000 feet in the sea. In his recent work, "Mikrogeologie," he has shown also that the most ancient of the fossil Infusoria, whether belonging to the Carboniferous or to the Silurian strata, belong to the same genera, and often to the same species, as those which actually exist at the present day.

"The minute grains of greensand," says this author, "which are characteristic of many rocks, have a different nature from the green earth often met with in concretionary masses. The former, from the *Glauconie* of the Paris limestone to the Azoic lower Silurian greensand near Petersburg, appear to consist of green opalescent casts of *Polythalamia*, composed of a hydrosilicate of iron. The cretaceous greensands of England contain, unmistakeably, these stony casts. In the Tertiary compact, limestone and nummulitic limestones, occur beautifully preserved specimens of *Quinqueloculina, Rotalia, Textularia, Grammostoma,* and *Alscolina.* In the lower Silurian greensand casts of detached cells of *Textularia* and *Nodosaria* have been found."

In the lakes of Sweden there are vast layers of iron oxide almost exclusively built up by animalcules. This kind of iron-stone is called lake-ore. In winter the Swedish peasant, who has but little to do in that season, makes holes in the ice of a lake, and with a long pole brings up mud, etc., until he comes upon an iron bank. A kind of sieve is then let down to extract the ore. One man can raise in this manner about one ton per diem.

Besides the excellent polishing material furnished by these infusorial deposits, Liebig has recently drawn attention to another application of which they are susceptible. His observations were made upon an infusorial deposit which constitutes the under soil of the commons or plains of Lünebourg, in Germany (Fig. 34); and he has shown that these microscopic remains, as well as those taken from several other localities, can be very easily converted into *silicate of potash* or *silicate of soda,* sometimes known as "*soluble glass.*" It was first ascertained by analysis that this infusorial earth contained 87 per cent. of pure silica. The following method was then adopted to convert it into silicate of soda :—148 lbs. of calcined carbonate of soda are dissolved in five times their weight of boiling water; to this is added a milk of lime pre-

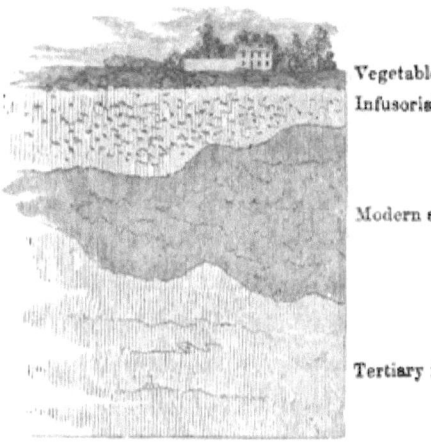

Fig. 34.—Infusorial Deposit, Lünebourg, Germany.

pared with 84 lbs. of quicklime. After boiling the mixture for ten minutes or a quarter of an hour, the alkaline liquid, which now contains caustic soda, is decanted off from the insoluble carbonate of lime, and evaporated in an iron vessel, until it has acquired a specific gravity of 1·15. At this moment 240 lbs. of the infusorial earth is added. The latter dissolves rapidly in the alkaline solution, and leaves scarcely any residue. If by any accident a smaller

s

quantity of infusorial earth than that prescribed be taken, the soluble glass obtained is too alkaline and very deliquescent.

Soluble glass, first discovered by the ingenious chemist, Fuchs, of Munich, is an alkaline silicate of potash or soda. It has been utilized in various ways, principally for protecting wood, linen, the scenery of theatres, panoramas, etc., from fire. Tissues steeped in it lose their faculty of burning with flame; if held in the fire they will consume slowly and without flaming, so that any such tissue being set on fire cannot communicate its combustibility to other substances near, and in nine cases out of ten it will not take fire at all.

These infusorial deposits, moreover, furnish very good material for the manufacture of window-glass, plate-glass, etc.; besides which they make an excellent mortar, and can be converted into filters, into moulds for casting iron, brass, or other metals. Add to this the use made of them as food and their polishing quality, and we shall see at a glance how much the remains of these invisible animalcules have been turned to account by man.

Chalk, also, which has innumerable uses—which is employed, for instance, to prepare mortar, cement, as a manure, as a polishing material for silver and gold, etc., for whitewashing, to prepare lime, etc.; chalk also appears to owe its origin to the remains

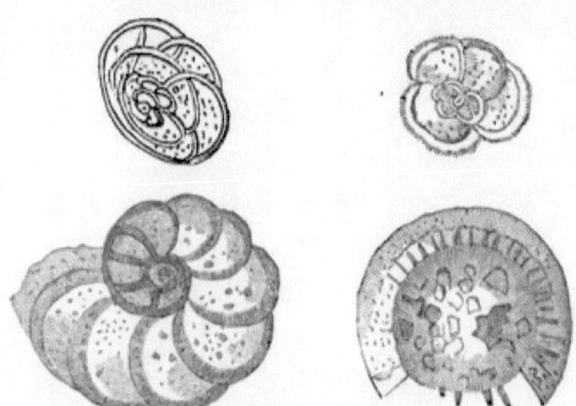

FIG. 36.
Foraminifera of the mud in which the Transatlantic Telegraph Cable lies (from nature, magnified 150 diameters).

of myriads of animalculæ, principally microscopic *Foraminifera* (Figs. 35 and 36).

These animalculæ, of which numerous species are still living, secrete a calcareous shell or covering,

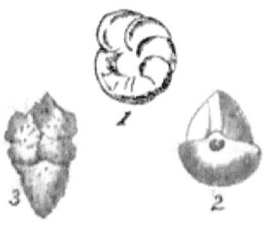

Fig. 35.—Foraminifera (magnified).
1. Rotalina.   2. Triloculina.   3. Sagrina.

similar to that of the siliceous infusoria. In spite of their minuteness, these shells offer several partitions or joints, which render them extremely beautiful; and as some of them resemble in miniature the *Nautilus* shell, some naturalists have been tempted to class them among the *Cephalopoda* mollusca, of which I have spoken; but very recent investigations invite us to place them as allies of *Infusoria*.

"These tiny shells," says Beudant, speaking of *Foraminifera*, "of which seven to eight hundred fossil species are already known, are found accumulated in immense masses in the terrestrial strata, and constitute of themselves enormous stratifications, of which the white chalk, and some of the tertiary limestones, furnish us with examples in every part of the world."

Traces more or less abundant of *Foraminifera* are to be found in the calcareous rocks of nearly every geological period; but it is towards the end of the secondary and at the commencement of the tertiary period, that the development of this group of fossils seems to have attained its maximum.

"Although there can be no reasonable doubt," says Dr. Carpenter, "that the formation of chalk is partly due to the disintegration of corals and larger shells, yet it cannot be questioned that in many localities a very large proportion of its mass has been formed by the slow accumulation of foraminiferous shells."

But the calcareous bed of the tertiary formations, known as *Nummulite limestone* (on account of the enormous quantity of Nummulite shells—larger Foraminifera—which it contains), is perhaps more interesting still. This Nummulitic limestone can be traced from the Pyrenees, through the Alps and Appenines, into Asia Minor, and further, through Northern Africa and Egypt, into Arabia, Persia, and Northern India; and thence, in all probability, through Thibet and China to the Pacific, covering very extensive areas, and attaining a thickness in some places of many thousand feet. Another tract of this remarkable strata is found in North America.

A similar deposit occurs in the Paris tertiary basin, and in that of Brussels; and it is not a little remarkable that the fine-grained and easily-worked limestone, which affords such an excellent material for the decorated buildings of the French capital, is almost entirely formed of accumulated masses of the minute shells of foraminiferous animalculæ. Even in this Nummulitic limestone, the matrix in which the Nummulites are imbedded is itself composed of the more minute *Foraminifera*, and of the broken and cemented fragments of the larger species.

It has often been remarked by chemists of repute, that, in whatever manner carbonate of lime was produced in the laboratory, nothing resembling chalk has ever been obtained. The mystery was solved when Ehrenberg showed us that this substance is almost entirely composed of fossil animalculæ, of which he counted as many as a million and a third in one cubic inch.

The manner in which these microscopic fossils may be rendered visible is thus:—On a plate of glass we place an extremely fine layer of chalk, which, when perfectly dry, is covered over with Canada balsam; and then, gently warming the whole, we observe with a magnifying power of two to three hundred diameters.

Seventy-one species of these *Foraminifera* were

soon detected in the white chalk, many of which may still be found living in the North Sea. It was also found that, in the chalk deposits of Southern Europe, the fossil animalculæ are beautifully preserved; whilst in the chalk of more northern latitudes, their shells are mostly found broken.

Microscopic vegetable forms, principally *Diatomaceæ*, abound also in the foraminiferous chalk, as in the other infusorial deposits of which I have spoken. Mr. E. O'Meara has lately found forty-two species of *Diatomaceæ* in the white chalk of Antrim, all of which are identical with living species.

When we consider the *time* that these immense deposits of animalculæ—such as the cliffs of Dover for instance—must have taken to accumulate, we can form no adequate idea of it, and we are once again reminded that *time* is the creation of man —that nature knows no *time!*

Chapter XI.

Sponges.

Remarks on Classification—Structure of a Sponge—Naturalists who have contributed to the history of Sponges—Chemical nature of Sponge—Interesting results—Spongia officinalis and S. usta—The Syrian toilet Sponge—Its high price—Other Sponges—Objects for the Aquarium—Spongilla fluviatilis and S. lucustris, or the fresh-water Sponges—Sponges common on the English Coast—Their use in Medicine—Sources of Iodine and Bromine—Flints and Agates, as owing their formation to Sponges—Petrified Sponges—Practical details on the toilet Sponge—Sponge Fishery and Sponge Markets.

SPONGES.

I HAVE placed *Sponges* in my last chapter, and in doing so I am apparently following the old zoological routine, which regards these singular beings as the last link of the animal chain—the link which joins the animal to the vegetable world; but this surely is not a fact! *Sponges* are evidently more closely allied to Polypes than to such animalcules as the *Monads*. Indeed, had it been practicable, I would willingly have condensed Polypes, Infusoria, and Sponges into one chapter. But the reason why *Infusoria* have been lately placed before *Sponges* by most zoologists appears to be, that as the former class becomes better known, and the organization of its species more thoroughly investigated by means of the powerful microscopes constructed at the present day, the complication of their structure excites astonishment, and, as we have already seen, many genera are being placed much higher in the series than the places which were formerly assigned to them. In the same way many *Infusoria* will probably, one day, be classed below *Sponges*. We must look upon a vast number of these microscopic beings as a group of animals under discussion. Proper places will be assigned to them as we become better acquainted with their organization. In the meanwhile it would be rash to attach too great an importance to the fact of my placing, in this work, *Infusoria* before *Sponges*, and *Polypes* before *Infusoria*, when, in a zoological point of view, they might, perhaps, for some years to come, be all jumbled into one chapter.

I stated in my last chapter, that *time* was a creation of man. It is equally evident that these zoological divisions are also the work of man, and as Nature knows no *time*, so also she knows no *division*. Nature is one harmonious whole, which man has cut up into sections in order to investigate this whole, piece by piece. One small piece generally suffices for many generations of human intellect!

Let us now see, in the fewest words possible, what a sponge is.

The sponge itself—*i. e.*, the substance we use as such—is composed of a horny flexible skeleton, forming a dense anastomosed tissue, in which numerous *pores* are seen. These are the openings of *canals* which traverse the sponge in all directions.

The canals are lined with a soft gelatinous animal matter, up to the opening of the pores themselves. The pores are strengthened, and probably kept open, by curious little needle-like bodies, called *spicula*, which are either siliceous or calcareous. Whilst the animal is alive, the water entering into the sponge by the pores circulates in the canals of the sponge, and is finally expelled through the larger openings, called *orifices* (or *oscula*), which are also observable on the surface, interspersed among the pores.

The currents thus observed are generated either by a ciliary apparatus existing in the gelatinous substance which lines the canals, or by capillarity.*

The currents from the orifices are best observed by placing a sponge, whilst alive, in a shallow dish of water, upon which a little powdered chalk has been thrown. The motions of the atoms of chalk will indicate precisely the direction of the currents. If the gelatinous matter which lines the canals be separated, by hot water, from the tissue or skeleton, the latter may be then examined under the microscope.

The gelatinous substance putrifies easily; it is of various colours, but principally yellowish-brown, and resembles the soft part of polypes.

* Consult on this Dutrochet, in the Memoirs cited on p. 271.

The *ova* of sponges are numerous irregularly-shaped granular bodies, endowed with vibrating cilia, by which they move. They issue at different periods from the gelatinous matter. These *ova* float in the water; moved about by the cilia which garnish their anterior extremity, they are carried on by the currents through the sponge, and are finally expelled through the larger orifices. They swim about freely in the water for a little while, and then fix themselves for ever to the rocks, and grow into new sponges. These *ova*, or moveable eggs, have frequently been taken for the animal (the sponge) itself.

The *spicula* are microscopic needles, sometimes straight, sometimes curved or star-shaped; others resemble the anchors of ships, etc., in form. When the spicula are siliceous, they are best seen after the sponge is burnt, on examining under the microscope the ash which is left.

Sponges with calcareous spicula are rather numerous on our coasts, and siliceous spicula are common in sponges of most latitudes.

It is almost entirely to English naturalists that we are indebted for the knowledge we possess of these curious organisms. Ellis was the first to establish the existence of currents of water passing constantly through the tissue of sponges. Dr. Grant, whilst confirming Ellis's observation, added so much valuable matter to the natural history of sponges, that his name has become European.*

The chemical nature of sponge is yet a problem to be solved, which may be said of many other animal products. However, something has been done, with a view to solve the difficulty, by Mulder, Crookewit, and Posselt. One of the most remarkable results obtained with regard to the chemical composition of the sponge is that arrived at by Crookewit, who, on analyzing a specimen of *Spongia officinalis*, discovered in it that peculiar substance called *fibroin*, which Mulder first extracted from the silk of the silkworm, as I stated in the proper place.

The analyses of this new product do not exactly agree, but they tend to show that fibroin contains 39 proportions of carbon, 62 of hydrogen, 12 of nitrogen, and 17 of oxygen. Besides this, sponge contains a certain proportion of phosphorus, of sulphur, and of iodine, which are combined, in some as yet unknown manner, with the fibroin. No albumine or gelatine have been found in sponges,

* See Ellis "On Corallines," and Grant "On Sponges," in "Edin. Phil. Journ." Also De Blainville, "Actinologie;" Lamouroux, "Genre des Polypes;" Dr. Fleming, "British Animals;" Dutrochet, "Mem. on the Spongilla," in his "Mem. pour servir à l'Hist. des Veg.," etc.; Bowerbank, in "Proceed. of the Geol. Soc.," and in "Microscopic Journ., 1841;" also "Brit. Ass. Rep., 1857."

as in silk. An elementary analysis of commercial sponge has given, in 100 parts—

Carbon	47·16
Hydrogen	6·31
Nitrogen	16·15
Oxygen	26·90
Iodine	1·08
Sulphur	0·50
Phosphorus	1·90
Bromine	traces
	100·00

Hence I conclude that the animal matter of sponge belongs to the group which contains fibrine, albumine, gelatine, etc., all of which give a per-centage of nitrogen resembling the above.

Winckler and Ragazzini have both shown that the ash obtained by the combustion of *Spongia usta* contains slight quantities of bromine.

These results are certainly not devoid of interest. Both Crookewit's and Posselt's analyses agree pretty well, and show that sponge contains rather more than 16 per cent. of nitrogen. It is, therefore, as rich in this element as the most valuable kinds of guano are.

The common sponge (*Spongia officinalis*, L.) is found abundantly in the Mediterranean, and will doubtless be *cultivated*, one of these days, by the

French upon the coasts of France and Algeria, though nothing of the sort has yet been attempted by them. It is imported at Liverpool from Turkey under the name of *Turkey sponge*, together with the West Indian, or *Bahamia sponge* (*Spongia usta*), a distinct species. The latter arrives in Liverpool from the Bahama Islands. The average importation to this seaport is about 135 cases per annum, each case containing about 500 sponges of various sizes, of which the average value is about 35*s*. per pound.

These two kinds of sponges form an important branch of commerce. The most prized for toilet purposes are the Syrian sponges. They are generally conical in shape, or sometimes hemispherical; the orifices of their internal canals are very small; they are hollow in the centre like a goblet, and their exterior possesses the softness of the finest velvet. I have seen some of these beautiful sponges selling in the Palais Royal, at Paris, for as much as 200 francs (£8) a piece. They were about five inches in diameter. Others, much smaller, were put up for sale at 50, 60, and 70 francs.

Besides the two species just named, there exist a number of others, some of which are common on our coasts, and astonish us by the beauty of their organization. The small parasitical sponges that cover the stalks of sea-weeds, or the larger varieties which cling to the rocks, well repay observation, and would form interesting objects for the aquarium. The same might be said of those two remarkable species of fresh-water sponges, *Spongilla fluviatilis* and *S. lacustris*. One of these species (*S. fluviatilis*) is not unfrequently met with in the ditches around Paris, and probably around London also. These *Spongilla* are green, and at first sight would be taken for vegetables. Mr. John Hogg has published, in the "Linnæan Transactions," some experiments made with a view of ascertaining the effect of light upon these fresh-water sponges. He has shown that they are influenced by it as vegetables are, and that their green colour depends upon their exposure to it. M. Dutrochet, in the memoir cited above, has studied minutely the organization of these fresh-water sponges.

To return to marine sponges, one of the most common of our indigenous species, *Spongia oculata*, or *Halichondria oculata* (Fig. 37), may be made to serve the same purposes as foreign sponges, save for the toilet; whilst *H. palmata*, *H. cervicornis*, *H. tubulosa*, *H. simulans*, etc., form beautiful specimens for the aquarium.

Carbonized sponge has been long used in medicine; its effects appear to depend upon the small quantity of iodine contained in it, of which, in

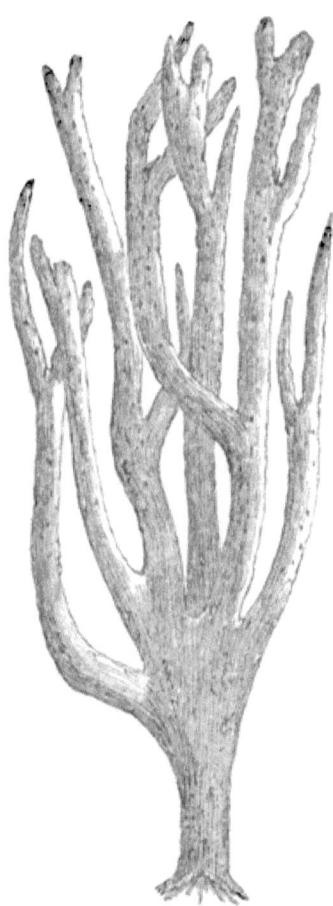

FIG. 37.
Spongia oculata (English sponge).

its natural state, this sponge contains about one per cent. It might, therefore, be a profitable speculation to extract this useful element from such sponges as *S. oculata* that abound on some of our English coasts. It is probable, also, that if all the different varieties of sponges, polypes, star-fish, etc., which are left to putrefy upon our shores, were properly collected, they would prove a valuable source of *iodine* and *bromine*, which are now, in spite of their high price, so much used in the chemical laboratory and by photographers. In places where sponges are abundant, the commoner sorts would prove useful to manure manufacturers, on account of the large per-centage of nitrogen they contain. They are soluble in strong acids, and also in alkaline solutions. It has been found the *S. tomentosa* (*S. urens*), which is common upon the coasts of England and North America, will raise blisters when rubbed upon the hand; and if previously dried in an oven, its stinging faculty is much increased.

According to Dr. J. S. Bowerbank, the flints of the chalk formation, and the beautiful moss agates which every one admires, are of spongeous origin; that is to say, have been formed by sponges which are now fossil. In fact, agates and flints are, according to this author, *petrified sponges*. It is indeed true that the polished section of a moss agate, or of certain flints, exhibits, in a beautiful manner, the structure of a sponge. Dr. Bowerbank's views on this subject are very clearly expressed in his paper read before the British Association in 1856, in which he brings forward numerous proofs of his theory, and to which I must refer my readers for the details. I agree with this author that sponges doubtless have, at various periods of the earth's history, largely contributed towards the formation of *agates* and *flints;* but it is evident, at the same time, that other siliceous deposits, such as those of fossil infusoria, etc., have a very different origin.

Flints generally contain numerous fossil infusoria, and indeed their formation has often been attributed to the remains of these animalculæ. At the same time, sponges appear to have contributed also to the formation of these curious stones; and here is a curious fact in relation to this:—In the south of Europe, the beds of marl which alternate with the white chalk consist of myriads of siliceous shells of *Infusoria* and *Diatomaceæ*, and flints are wanting; whilst in the north of Europe the reverse is found to be the case—beds of flint are met with, and marls with infusoria are wanting.

Flints not only show beautifully-preserved remains of sponges, but also those of polypes, such as *Alcyonia*, etc., *Echinia*, and other marine organisms, even molluscous shells or their impressions, numerous infusoria, and star-like microscopic objects, which have been taken for fossil animalculæ, and termed *Xanthidia*, but which are probably the *spicula* of fossil sponges.

The colour of flints, agates, etc., is owing to organic matter, and is consequently destroyed by heat. When calcined and ground to powder, flints are used to manufacture the finer sorts of pottery, and which is termed *flint-glass*. Before the invention of percussion-caps, gun-flints were in general use. It is a curious fact that sponges, one of the softest of animal structures, should have contributed so much to form one of the hardest of mineral substances, and that men have made war and slaughtered many thousands of their fellow-creatures by means of sponges and infusoria!

Flints also form an excellent building material, because they give a firm hold to the mortar, and resist every vicissitude of weather. The counties of Kent, Essex, Suffolk, Norfolk, etc., afford examples of many substantial constructions in flint masonry.

The uses of agates, for brooches, rings, seals, etc., are too well known to need mention here.

To return now to the toilet sponge, which constitutes such an important article of commerce, and about which I will add a few practical details.

The exact time required for the growth of the rigid portion or skeleton of the sponge, and the duration of this skeleton, is not known with accuracy; but it appears, from recent investigations, that beds of sponges spring up and increase rapidly where they were not before observed, and that a period of *two years* is generally sufficient to renew the crop of sponges on rocks that have been laid almost bare by the sponge fisheries. It has also been asserted that of all the numerous varieties of sponge already known, that which possesses the most precious qualities for the toilet grows in the Mediterranean. The places where its growth is most abundant are in the Grecian archipelago, the coasts of Syria and those of Barbary. The sponge fishery there is a profitable trade, and although perfectly free, it is scarcely practised by any others than the Greeks and the inhabitants of the shores on which sponges grow luxuriantly.

A strong constitution and a certain intrepidity being required, the sponge fishery is almost completely monopolized by the Greek and Arabian divers.

The coarser varieties of sponge are brought up from a comparatively slight depth, but for the soft, delicate varieties it is sometimes necessary to dive down thirty fathoms or more.

As soon as they are taken from the water, the

sponges undergo a very essential operation. They are placed in large round shallow holes dug in the sand of the coast, and filled with water, where they are trampled upon by the men until they are divested of their gelatinous animal matter and other impurities.

Beyrouth, Lattakiek, and above all Tripoli, are the most important sponge markets. Strangers arrive at Tripoli—where the fine landscape recalls the beautiful environs of Eden, which is only eight leagues distant—from all parts of the Levant, from every point of the Mediterranean, and even from Paris. Nothing can be more curious than this *mélange* of people of every nation drawn to one spot during the sponge season, every individual striving to outdo his neighbour, and competing to his utmost with the commercial dexterity of the keen Greek sponge merchants.

The market at Tripoli is held about the middle of September, a period at which the sponge fishery, like our work, draws to an end.

Note.—Since this volume was written, I find in the "Intellectual Observer" for January, 1864, a valuable article upon the Tinnevelly Pearl Banks, by Clements R. Markham, Esq., in which the author, whose views coincide perfectly with my own, gives much interesting information regarding the Asiatic Pearl Fisheries, showing the absolute necessity of establishing a more rigorous method and a proper *cultivation* of the pearl-oyster, based upon scientific observation, in order to reform the present unsatisfactory state of these fisheries.

THE END.

LIST OF WORKS AND PHILOSOPHICAL PAPERS
By DR. T. L. PHIPSON, F.C.S. LOND.,

Late of the University of Bruxelles; Member of the Chemical Society of Paris; Laureate of the Dutch Society of Sciences; Corr. Memb. of the Belgian Entomological Society, the Pharmaceutical Society of Antwerp, the Society of Medical and Natural Sciences of Bruxelles, the Society of Sciences of Strasburg, etc., one of the Editors of "Le Cosmos," etc., etc.

---

1. The Utilization of Minute Life. 8vo. London, 1864. Groombridge and Sons.
2. Phosphorescence; or, the Emission of Light by Minerals, Plants, and Animals. 8vo. London, 1862. Reeve and Co.
3. La Force Catalytique, Etudes sur les Phenomènes de Contact (Prize Essay, Dutch Society of Sciences). 4to. Harlem, 1858. Loosjes.
4. Le Préparateur-Photographe, traité de Chimie à l'usage des Photographes, etc. 8vo. Paris, 1864. Leiber.
5. Essay on the Uses of Salt in Agriculture (Prize Essay). London, 1863. Simpkin.
6. Mémoire sur le Fécule et les Substances qui peuvent la remplacer dans l'Industrie. Bruxelles, 1854. Tircher.
7. Recherches nouvelles sur le Phosphore. Bruxelles, 1855. Tircher.
8. Essai sur les Animaux Domestiques des Ordres Inférieurs. Paris, 1857. Leiber.

In the *Journal of the Chemical Society*, 1862 to 1864.

1. On the Transformations of Citric, Butyric, and Valerianic Acids. 1862.
2. On Sombrerite, a new mineral. 1862.
3. On the Bicarbonate of Ammonia of the Chinca Isles. 1863.
4. On Vanadium Ochre, and other sources of Vanadic Acid. 1863.

In the *Proceedings of the Royal Society*, 1863 to 1864.

1. Researches on several Mineral Substances, including determining their Analysis, etc.
2. On Magnesium.
3. Note on the Variations of Density produced by Heat in Mineral Substances.

In *Comptes-Rendus de l'Académie des Sciences de Paris*, 1856 to 1863.

1. De l'Action des Corps Organiques sur l'Oxygène. 1856.
2. Sur la Production de la Mannite par les Plantes Marines. 1856.
3. Sur une Nouvelle Roche de Formation Récente, etc. (1857 and 1860, two notes).
4. Sur quelques Phenomènes Météorologiques observés sur le littoral de la Flandre. 1857.
5. Notes sur les Térédo Fossiles. 1857.
6. Sur une Pluie sans Nuages observée à Paris. 1857.
7. Sur la Putréfaction à 35 degrés sous zéro. 1857.
8. Action de la Santonine sur la Vue. 1859.
9. Sur la Présence de l'Aniline dans certains Champignons. 1860.
10. Sur une Pluie de foin observée à Londres. 1861.
11. Sur quelques cas nouveaux de Phosphorescence par la Chaleur. 1860.
12. Sur la Matière Phosphorescente de la Raie. 1860.
13. Sur un Oxide d'Antimoine natif de Borneo. 1861.
14. Sur le Tinkalzite de Perou. 1861.
15. Sur un Brouillard sec à Londres. 1861.
16. Sur la Couleur des Feuilles. 1858.
17. Sur le Soufre Arsénifère des Solfatares de Naples, et sur la Préparation du Sélénium. 1862.
18. Sur l'acide Manganique. 1860.
19. Sur un Oligiste de l'Époque Dévonien et sur une Matière Organique qu'il contient. 1861.

In the *Chemical News and Journal of Physical Science*, 1860 to 1864.

1. On a new Sulphide of Chromium. 1861.
2. Note on Fluorine. 1861.
3. On a new Colouring-matter. 1861.
4. Experiments and Observations on the part played by Oxygen in Eremacausis and Fermentations. 1863.
5. On the presence of Xanthic Oxide in Guanos containing no Uric Acid. 1862.
6. Analysis of the Diluvial Soil of Brabant, etc. 1862.
7. On the Argentiferous Gossan of Cornwall. 1862.
8. Analysis of a Specimen of Fossil Wood from the Green-sand of the Isle of Wight. 1862.
9. Composition of a peculiar substance which exudes from a Tertiary rock in Australia. 1862.

10. On Native Zinc and Native Tin. 1862.
11. On Crystallized Platinum. 1862.
12. Artificial formation of Populine. 1862.
13. On a new Harmonica Chymica. 1862.
14. On Musical Sounds produced by Carbon. 1863.
15. Determination of Specific Gravity of Mineral Substances. 1862.
16. On Zinc Green. 1863.
17. On a new method of Measuring the Chemical Action of the Sun's Rays. 1863.
18. Note on Vegetable Ivory. 1863.
19. On the constant increase of Organic Matter in Cultivated Soils. 1863.
20. On the Composition of Gas-refuse. 1863.
21. Potabilisation of Sea-water by the Electric Current. 1863.

In the *Journal de Médecine et de Pharmacologie de Bruxelles*, from 1854 to 1862 inclusively.

1. Expériences et Observations sur la Présence de l'Ammoniaque dans la Respiration. 1856.
2. Action de l'Acide Sulfurique sur le Zinc et le Fer. 1858 (two papers).
3. Quelques mots sur les Modifications Allotropiques des gaz. 1855.
4. Sur l'Oxygène Allotropique, etc. 1856.
5. Encore quelques mots sur l'Ozone, etc. 1856.
6. Sur les Produits de la Distillation sèche des Matières fécales. 1857.
7. Sur le Vert de Zinc. 1857.
8. Sur les grenats Naturels et Artificiels. 1857.
9. Analyse d'un Mélange Gazeux Contenant du l'Oxygène. 1856.
10. Sur les Bolets bleuissants, Etude de la Formation des Matières Colorantes chez les Champignons. 1860.
11. Protoctista ou la Science de la Création aux points de vue de la Chimie et de la Physiologie. 1861.
12. Analyses de quelques Substances Minérales. 1862.
13. Sur la Forme Crystalline du Charbon. 1859.
14. Sur une nouvelle Théorie d'Ethérification. 1855.
15. Sur le Fluorure de Potassium. 1858.
16. Sur les Oxalates de Fer. 1861.
17. Sur la Théorie Electro-Chimique. 1856.

MISCELLANEOUS WRITINGS.

In the *Geologist*, Vols. i. and ii., 1858 to July 1859.
Foreign Correspondence. 19 Papers.

In the *Intellectual Observer*, 1864.
Vanadic Acid.   The Phosphates used in Agriculture.

In the *Popular Science Review*, 1863 to 1864.
Anæsthetics.   The Aniline Dyes.

In *Macmillan's Magazine*, 1862 to 1864.
Electricity at Work.   Gold, its Chemistry and Mineralogy.
The Chemistry of the Sea.   The Movements of Plants.

In the *Cosmos*, Paris, 1856 to 1864. 18 vols.
Reviews, Miscellaneous Articles, and English Correspondence.

In the *Moniteur de la Photagraphie*, Paris, 1861 to 1864. 4 vols.
English Correspondence.

In the *Technologist*, 1861, and *Photographic News*, 1861.
On a New Process of Photography without Silver.

In the *Progrès par la Science*, Bruxelles, 1864.
Etudes de Chimie Agricole.

NEW BOOKS AND NEW EDITIONS
PUBLISHED BY
GROOMBRIDGE AND SONS.

New Work by the Author of "The Heir of Redclyffe," etc.
Fcap. 8vo, printed on toned paper, with 36 Initial Letters and other Illustrations, cloth gilt antique, price 5s.,

THE WARS OF WAPSBURGH.
BY THE AUTHOR OF "THE HEIR OF REDCLYFFE," Etc.

"Quite worthy of the Authoress of 'The Heir of Redclyffe.'"—*Notes and Queries.*

"A charmingly fanciful story. Every one who hates wasps ought to read this book and be converted. An atmosphere of kindly human interest is thrown over the whole tale, which is lighted up by the play of a beautiful fancy."—*Reader.*

"This very nicely printed book we can recommend most heartily. The allegory is well carried out."—*Standard.*

THE OLD BUSHMAN IN LAPLAND.
AT ALL THE LIBRARIES. Ready this day, post 8vo, cloth gilt, price 10s. 6d.,

A SPRING AND SUMMER IN LAPLAND,
WITH NOTES ON THE FAUNA OF LULEÄ LAPMARK.
By AN OLD BUSHMAN,
Author of "Bush Wanderings in Australia."

"**We trust** that the Old Bushman's book may send many a true naturalist, and many a holiday maker too, to the country that his book so well describes."—*Reader.*

"As a book for general reading, 'A Spring and Summer in Lapland' will be found one of the pleasantest of the season."—*Intellectual Observer.*

"The description of his being lost for nine hours at night in a snow storm is distressingly vivid; we doubt whether Defoe or George Eliot ever wrote anything finer in point of physical and psychological description. There is an agonizing simplicity, a depth, force, and truth of detail which could hardly be surpassed, because every touch is in the nature of the thing."—*Spectator.*

"A volume which will be acceptable to the ornithologist and the sportsman."—*Observer.*

"His notes abound in information."—*Sun.*

"It was a good thought that took the Old Bushman on a hunting and naturalist's mission to Lapland. His volume, telling of the natural features of this district, and of the many animals that are almost its only inhabitants, is more full of new and solid matter than the majority of travel books, and therefore has greater claims on the attention of men of science."—*Examiner.*

"There is no need to praise such books as this, which will attract and delight many readers. No review does it justice."—*Standard.*

"These notes on Lapland will be very acceptable to lovers of Natural History, and particularly so to students of ornithology."—*Notes and Queries.*

"As a chronicler of these facts, the 'Old Bushman' is strikingly conscientious. He has recorded nothing which did not come under his own personal observation."—*Saturday Review.*

"Independently of the valuable zoological information, and the useful hints to sportsmen which are to be found in its pages, the book abounds with illustrative anecdotes, incidents of northern travel, the Author's account of his being lost in the snow, and many other details of his experiences, render the Old Bushman's work thoroughly worth reading."—*Athenæum.*

"What can the British sportsman require more? Let him at once start to spend the 'Spring and Summer in Lapland,' take the pleasant, carefully-written volume of the 'Old Bushman' in his pocket as—next to a circular note—his best travelling companion."—*Daily News.*

London: GROOMBRIDGE & SONS, 5, Paternoster Row.

NEW BOOKS AND NEW EDITIONS

Dedicated, by permission, to H.R.H. the Duke of Cambridge.
Fcap. 8vo, cloth, price 5s.,

CURIOSITIES OF WAR AND MILITARY STUDIES.

By THOMAS CARTER, Author of "Medals of the British Army."

This Soldiers' Book and Military Compendium is full of the most interesting anecdotes and incidents relating to various regiments, and contains the origin of every infantry corps, together with names and heroic deeds of the recipients of the Victoria Cross, arranged regimentally.

Complete in Three Volumes 8vo, price 7s. 6d. each, cloth gilt, with Fac-simile Illustrations of the Medals and Ribbons in colours,

MEDALS OF THE BRITISH ARMY, AND HOW THEY WERE WON.

By THOMAS CARTER, Author of "Curiosities of War and Military Studies."

FIRST VOLUME.—THE CRIMEAN CAMPAIGN.

Dedicated to Major-General the Hon. Sir James Yorke Scarlett, K.C.B., Adjutant-General of Her Majesty's Forces. With Fac-simile Illustrations in Colours of

The Crimean Medal.	Sardinian War Medal.	Medal for Distinguished
French War Medal.	Turkish War Medal.	Conduct in the Field
	Victoria Cross.	

SECOND VOLUME.—EGYPT, PENINSULA, WATERLOO, & SOUTH AFRICA.

Dedicated to His Grace the Duke of Richmond, K.G., etc., etc. With Fac-simile Illustrations in Colours of

The Gold Cross.	Talavera Medal.	Medal for Meritorious
The War Medal.	Waterloo Medal.	Service.
Turkish Medal for Egypt.		The Cape Medal.

THIRD VOLUME.—INDIA, CHINA, AND PERSIA.

Dedicated to General Lord Clyde, G.C.B., etc., etc. With Fac-simile Illustrations in Colours of

The Indian War Medal.	The China Medal.
Seringapatam Medal.	The Second Jellalabad Medal.
The Sutlej and Punjab Medals.	The Maharajapoor Star.
The Ghuznee Medal.	Medal for the Second Burmese War.
Indian Mutiny Medal.	And other Illustrations.

CRITICAL NOTICES.

"Such a book has a national interest, and should have been brought out at the public expense. Though this has not been done, a copy will doubtless find its way into every regimental library, for it is a common record of glory, in which every regiment is enrolled."

"We cannot estimate the good which may be effected by its pages; for soldiers will be inspired to emulate the deeds which they see so perpetuated, and which our medals embellish. The book is particularly suited for this object, as it contains no strictures on authority, or anything to awaken bad feelings. At the same time, it will serve as an admirable manual for officers who wish to have a knowledge of our military history, which is here brought into a nutshell, though no deed of note is omitted."—*United Service Magazine.*

"The handsome book before us is one Muster Roll of England's brave men."—*London Review.*

"The illustrations are particularly well done, and we may say of the work generally that it is worthy of the subject."—*Russell's Army and Navy Gazette.*

"We must here state that an interesting and officially accurate description of each battle, etc., as well as a description of the medals, is given in each case; and the pages of the work are adorned by beautifully executed chromo-lithographs of the medals, clasps, ribbons, etc. The fact of Mr. Carter's position in the military executive branch of the War Department, and the sanctioned dedications of the several divisions of the work to the late Duke of Richmond, General Lord Clyde, and the Adjutant-General, must greatly add to its value as a book of reference, and as authority upon the subject treated."—*The Art Journal.*

The Illustrations include the several Medals issued to the Army, from the earliest one granted for Cromwell's victory of Dunbar to the most recent for the Indian Mutiny and Services in China. Fac-similes of the Medals and Ribbons in colours, gold, silver, and bronze, are also given of those conferred upon our troops for the Crimean campaign by the Emperor of the French, King of Sardinia, and the Sultan of Turkey, together with the VICTORIA CROSS.—The names and deeds of the Recipients are embodied in the Text.

London: GROOMBRIDGE & SONS, 5, Paternoster Row.

NEW BOOKS AND NEW EDITIONS.

THE WORKS OF GRACE AGUILAR.

HOME INFLUENCE. A Tale for Mothers and Daughters.
Fcap. 8vo, Illustrated, cloth gilt, 5s.

THE MOTHER'S RECOMPENSE. A Sequel to Home Influence. With a Portrait of the Author, and other Illustrations. Fcap. 8vo, cloth gilt, 6s.

WOMAN'S FRIENDSHIP. A Story of Domestic Life.
Fcap. 8vo, Illustrated, cloth gilt, 5s.

THE VALE OF CEDARS; or, The Martyr. Fcap. 8vo, Illustrated, cloth gilt, 5s.

THE DAYS OF BRUCE. A Story from Scottish History.
Fcap. 8vo, Illustrated, cloth gilt, 6s.

HOME SCENES AND HEART STUDIES. Tales.
Fcap. 8vo, with Frontispiece, cloth gilt, 5s.

THE WOMEN OF ISRAEL. Characters and Sketches from the Holy Scriptures. Two vols., fcap. 8vo, cloth gilt, 10s.

CHEAP SERIES OF POPULAR BOOKS.
Price 2s. 6d. each.

UNDER BOW BELLS: a City Book for all Readers.
By JOHN HOLLINGSHEAD. Price Half-a-Crown.

ODD JOURNEYS. By JOHN HOLLINGSHEAD. Price Half-a-Crown.

WAYS OF LIFE. By JOHN HOLLINGSHEAD. Price Half-a-Crown.

UNDERGROUND LONDON. By JOHN HOLLINGSHEAD.
Price Half-a-Crown.

SELF AND SELF-SACRIFICE; or, Nelly's Story. By ANNA LISLE. Price Half-a-Crown.

ALMOST; or, Crooked Ways. By ANNA LISLE. Price Half-a-Crown.

QUICKSANDS. A Tale. By ANNA LISLE. Price Half-a-Crown.

PICTURES IN A MIRROR. By W. MOY THOMAS.
Price Half-a-Crown.

LYDIA: A Woman's Book. By Mrs. NEWTON CROSLAND.
Price Half-a-Crown.

A FEW OUT OF THOUSANDS: their Sayings and Doings. By AUGUSTA JOHNSTONE. Price Half-a-Crown.

FOOTSTEPS TO FAME: A Book to Open other Books.
By HAIN FRISWELL. Price Half-a-Crown.

LEAVES FROM A FAMILY JOURNAL. By EMILE SOUVESTRE. Price Half-a-Crown.

London: GROOMBRIDGE & SONS, 5, Paternoster Row.

NEW BOOKS AND NEW EDITIONS.

GROOMBRIDGE'S SHILLING GIFT BOOKS.
SELECTED FROM THE MAGNET STORIES.
Elegantly bound, cloth gilt, for Presentation.

UNION JACK, and other Stories. By Mrs. S. C. HALL. Containing "Union Jack," "Mamma Milly," "Fanny's Fancies." Illustrated with Fifteen Engravings, cloth gilt, 1s.

THE TOWN OF TOYS, and other Stories. By SARA WOOD. Containing "The Town of Toys," "Hope Deferred," "The Merivales." Illustrated with Fifteen Engravings, cloth gilt, 1s.

NO-MAN'S LAND, and other Stories. By THOMAS MILLER. Containing "No-Man's Land," "Sweet Spring Time," "Golden Autumn." Illustrated with Fifteen Engravings, cloth gilt, 1s.

THE SEA SPLEENWORT, and other Stories. By the Author of "The Heir of Redclyffe," etc. Containing "The Sea Spleenwort," "The Mice at Play," "The Strayed Falcon." Illustrated with Fifteen Engravings, cloth gilt, 1s.

LOTTIE'S HALF-SOVEREIGN, and other Stories. By Mrs. RUSSELL GRAY. Containing "Lottie's Half-Sovereign," "Music from the Mountain," "My Longest Walk." Illustrated with Fifteen Engravings, cl. gt., 1s.

THE SHEPHERD LORD, and other Stories. By JULIA CORNER. Containing "The Shepherd Lord," "Hereward the Brave," "Caldas: a Story of Stonehenge." Illustrated with Fifteen Engravings, cloth gilt, 1s.

THE CAPTIVE'S DAUGHTER, and other Stories. By W. HEARD HILLYARD. Containing "The Captive's Daughter," "The Little Trapper," "The Planter's Son." Illustrated with Fifteen Engravings, cloth gilt, 1s.

THE ORPHANS OF ELFHOLM, and other Stories. By FRANCES BROWNE. Containing "The Orphans of Elfholm," "The Poor Cousin," "The Young Foresters." Illustrated with Fifteen Engravings, cloth gilt, 1s.

PLAYS FOR HOME ACTING AND YOUNG PERFORMERS.
BY JULIA CORNER.

THE KING AND THE TROUBADOUR. A Play for Home Acting and Young Performers. With a Coloured Frontispiece and other Illustrations. Imp. 16mo, gilt edges, 1s.

SLEEPING BEAUTY. A Play for Home Acting and Young Performers. With a Coloured Frontispiece and other Illustrations. Imp. 16mo, gilt edges, 1s.

GIFT BOOKS FOR THE YOUNG.

THE HISTORY OF A SHIP FROM HER CRADLE TO HER GRAVE. By GRANDPA BEN. Illustrated with more than One Hundred Engravings. Imp. 18mo, cloth gilt, 3s.

*** A most attractive book for boys is "The History of a Ship from her Cradle to her Grave." A perfect description of a Ship in all her parts, from the keel to the topsail; a book to be read and remembered, written by an Author skilled in nautical matters, well read in nautical history, and deeply acquainted with the life of a sailor.

VESSELS AND VOYAGES. A Book for Boys. By UNCLE GEORGE. Illustrated with Twenty Engravings, 16mo, cloth gilt, 1s. 6d.

OUT AND ABOUT. A Boy's Adventures. By HAIN FRISWELL. Author of "Footsteps to Fame." Illustrated by GEORGE CRUIKSHANK. Fcap. 8vo, cloth gilt, 3s. 6d.

CHRONICLES OF AN OLD OAK; or, Sketches of English Life and History. By EMILY TAYLOR, Author of "The Boy and the Birds," etc. With full-page Illustrations and Vignettes. Imp. 16mo, cloth gilt, 3s. 6d.

CHILDREN OF OTHER LANDS. Some Play-time Tales for Children of England. By SARA WOOD. Illustrated with Frontispiece and Vignettes. Imp. 16mo, cloth gilt, 3s. 6d.

London: GROOMBRIDGE & SONS, 5, Paternoster Row.

NEW BOOKS AND NEW EDITIONS.

Ready this day, post 8vo, cloth gilt, price 5s.,

ENGLAND'S WORKSHOPS.

BY

DR. G. L. M. STRAUSS.
C. W. QUIN, F.C.S.
JOHN C. BROUGH.

THOMAS ARCHER.
W. B. TEGETMEIER.
W. J. PROWSE.

CONTENTS.

METAL WORKSHOPS.

Mr. Gillott's Steel Pen Manufactory, at Birmingham.
The Gas Branch and Chandelier Manufactory of Messrs. Stroud & Co., at Birmingham.
Mr. Charles Reeves' Smallarms Factory.
Weighbridges at the Albion Works of Messrs. Pooley & Son, Liverpool.
Coal and Iron at Coalbrookdale.
A Canister Maker's.
Fitzroy Zinc and Galvanized Iron Works.
Brass Founding.
Brass-Foundry and Tube Works of William Tonks & Sons, Moseley Street, Birmingham.
German Silver.
Wrought Iron.
Dartmouth Works, Birmingham.
Tin Plate.

Electrum, Albata, and Virginian Plate Manufactory, and Electroplating Works of John Yates & Sons, Pritchett Street and Coleshill Street, Birmingham.
Tin-plate, Japanning, and Papier Maché Works, of Loveridge & Schoolbred, Merridale Street, Wolverhampton.
Edge Tools.
Messrs. John Yates & Co.'s Manufactory of Edge Tools, Exchange Works, Aston, and Pritchett Street Works, Birmingham.
Agricultural Implements and Machines.—A Day at the Orwell Works, Ipswich.
Sheffield Steel-ware.
The Queen's Plate and Cutlery Works of Messrs. Mappin Brothers, Baker's Hill.
Locks and Keys.

CHEMICAL WORKSHOPS.

The Great Chemical Works of Messrs. Chance Brothers & Co., at Oldbury.
A Visit to Messrs. Howard & Sons' Quinine, Borax, and Tartaric Acid Works, Stratford.
A Visit to Messrs. Davy & Macmurdo's Chemical Works, at Bermondsey and Upper Thames Street.

A Visit to Messrs. Huskisson & Sons' Chemical Factory.
Perfumes and Perfumery.—A Visit to Messrs. Piesse & Lubin's Laboratory of Flowers.
Messrs. Cliff & Co.'s Chemical Stoneware Works, Lambeth.

GLASS WORKSHOPS.

The Glass Works of Messrs. Chance Brothers & Co., Spon Lane, near Birmingham.

The Glass Manufactory of Messrs. Defries, in Houndsditch.

PROVISION AND SUPPLY WORKSHOPS.

Price's Patent Candle Company, Sherwood Works, Battersea.
Visit to the Lambeth Marsh Candle Works.
Visit to a Wax Vesta and Lucifer Match Factory.
The Bathgate Paraffin Oil Works.
Visit to a Provision, Cigar, and Wholesale Grocery Establishment.

Visit to a Tobacco Manufactory.
Paper Bags.
Mustard and Starch.—A Day at the Carrow Works, Norwich.
Messrs. Hill, Evans, & Co.'s Vinegar Works, at Worcester.
Messrs. Allsopp's Pale Ale Brewery, Burton-on-Trent.

DOMESTIC WORKSHOPS.

The Boar's Head Cotton Mills, Messrs. Evans & Co., Darley.

The Gray's Inn Pianoforte Manufactory.

"A work of great merit, and one which cannot be too highly commended."—*Observer.*

"No fairy pantomime contains more marvellous transformations than those which are related in the magic records of this book."—*Sun.*

"The idea of this book is good, and has, so far as we have been able to observe, been very well carried out."—*Reader.*

"An excellent and interesting account of the processes by which some of our commonest articles of utility are produced, and the wealth, science, and power employed in their production."—*Notes and Queries.*

London: GROOMBRIDGE & SONS, 5, Paternoster Row.

Price 2s. 6d., illustrated with full-page Engravings and Vignettes, appropriately bound in magenta cloth gilt—for PRESENTATION.

THE MAGNET STORIES,

FOR SUMMER DAYS AND WINTER NIGHTS.

CONTENTS OF THE FIRST VOLUME.

When we were Young. By the Author of "A Trap to Catch a Sunbeam."
Lottie's Half-Sovereign. By Mrs. Russell Gray.
Mamma Milly. By Mrs. S. C. Hall.
Havering Hall. By G. E. Sargent.
Blind Ursula. By Mrs. Webb (Author of "Naomi").
The Clockmaker of Lyons. By E. M. Piper.
The Mice at Play. By the Author of "The Heir of Redclyffe."

CONTENTS OF THE SECOND VOLUME.

Union Jack. By Mrs. S. C. Hall.
The Captive's Daughter. By W. Heard Hillyard.
Dear Charlotte's Boys. By Emily Taylor.
The Town of Toys. By Sara Wood.
Not Clever. By Frances M. Wilbraham.
Sea-Shell Island. By G. E. Sargent.
The Pedlar's Hoard. By Mark Lemon.

CONTENTS OF THE THIRD VOLUME.

The Story of Nelson. By W. H. G. Kingston.
Lost in the Wood. By Mrs. Alex. Gilchrist.
The Shepherd Lord. By Julia Corner.
Cousin Davis's Wards. By Margaret Howitt.
Hope Deferred. By Sara Wood.
Which was the Bravest? By L. A. Hall.
The Strayed Falcon. By the Author of "The Heir of Redclyffe," etc.

CONTENTS OF THE FOURTH VOLUME.

The Angel Unawares. By Mary Howitt.
The Little Trapper. By W. Heard Hillyard.
Music from the Mountains. By Mrs. Russell Gray.
Hereward the Brave. By Julia Corner.
Deaf and Dumb. By Mrs. Webb (Author of "Naomi").
An Adventure on the Black Mountain. By F. M. Wilbraham.
No-Man's Land. By Thomas Miller.

CONTENTS OF THE FIFTH VOLUME.

Coraline. By the Author of "A Trap to Catch a Sunbeam."
The Orphans of Elfholm. By Frances Browne.
The Story of a Pebble. By L. A. Hall.
The Sea Spleenwort. By the Author of "The Heir of Redclyffe," etc., etc.
The Christmas Rose. By H. J. Wood.
Ellis Gordon of Bolton Farm. By Emily Taylor.
The Grateful Indian. By W. H. G. Kingston.

CONTENTS OF THE SIXTH VOLUME.

Fanny's Fancies. By Mrs. S. C. Hall.
Sweet Spring Time. By Thos. Miller.
Caldas, a Story of Stonehenge. By Julia Corner.
The Poor Cousin. By Frances Browne.
The Planter's Son. By W. Heard Hillyard.
The Merivales. By Sara Wood.
Peter Drake's Dream. By Francis Freeling Broderip.

* Each Volume is so bound as to form a distinct and handsome Gift Book. The Six Volumes contain 42 Stories, either of which may be had separately, price 3d. each.

OPINIONS OF THE PRESS.

"The best guarantee for their excellence is that they are written by our best authors. The stories in the series are *all* excellent, and very well illustrated."—*Art Journal.*

"Each of these volumes makes an attractive present."—*City Press.*

"We recommend THE MAGNET STORIES strongly to those who wish to provide amusing, instructive, and, at the same time, really good reading for the young, at a trifling outlay."—*Eng. Churchman.*

"THE MAGNET STORIES we believe to be the best collection of children's books ever published."—*Brighton Gazette.*

"Published every month, and costing only Threepence, almost every mother in the land may with them delight the juveniles, among whom we know them to be huge favourites."—*Standard.*

London: GROOMBRIDGE & SONS, 5, Paternoster Row.

NEW BOOKS AND NEW EDITIONS.

Ready this day, illustrated with Coloured Plates, cloth gilt, price 7s. 6d.

MICROSCOPE TEACHINGS.

Descriptions of various Objects of especial Interest and Beauty adapted for Microscopic Observation. Illustrated by the Author's Original Drawings. With Directions for the Arrangement of a Microscope, and the collection and Mounting of Objects.

By the HON. MRS. WARD, Author of "Telescope Teachings."

ILLUSTRATED WITH SIXTEEN COLOURED PLATES,
DELINEATING

Wings of Earwig.	Hairs of Insects.
———— Wasp.	Down of Birds.
———— Beetles.	Structure of a Feather.
Wings and Scales of Moths and Butterflies.	Structure of the Crystalline Lens.
Scales of Beetles and Fishes.	Eyes of Dragon Fly.
Hair of Mouse.	———— Cricket.
———— Rabbit.	———— Lobster.
———— Cat.	Feet of Insects.
———— Otter.	Petal of Geranium.
———— Bat.	Pollen of Flowers.
———— Horse.	Seed Vessels of Ferns.
———— Deer.	Section of Limestone.
Human Hair.	Animalculæ.
Wool.	Circulation of the Blood in Fish, Frog, Newt, and Bat.

"The Telescope and Microscope unfold the same truths, and in all alike, 'while we explore and study, we feel a new sense of the unfailing power and infinite wisdom of the Great Creator, whose mercies are over all his works.'—So says Mrs. WARD in the preface to her charming volume, *Microscope Teachings*, and few books could better enforce the lesson. It is the best of several popular guides to the use of the Microscope."—*Examiner.*

"Nothing could be better than the way in which the drawing and colouring are done in the sixteen large coloured plates, each of them containing on an average half-a-dozen beautiful objects, and the text is worthy of the plates."—*Standard.*

"The work is altogether a most charming and appropriate introduction to the study of the Microscope."—*Athenæum.*

"The drawings are accurate and original, and we are happy to welcome Mrs. WARD amongst the earnest students of microscopic science. To our readers we cannot give better advice than, become purchasers of the book—they will not regret the outlay."—*Electrician.*

SHIRLEY HIBBERD'S WORKS.

RUSTIC ADORNMENTS FOR HOMES OF TASTE. With Recreations for Town Folk in the Study and Imitation of Nature. With Illustrations, Plain and Coloured. Second Edition, crown 8vo, cloth gilt, 14s.

THE BOOK OF THE AQUARIUM: Instructions on the Formation, Stocking, and Management, in all Seasons, of Collections of Marine and River Animals and Plants. New Edition, revised and additionally Illustrated. Fcap. 8vo, cloth gilt, 3s. 6d.

THE BOOK OF THE FRESH-WATER AQUARIUM. Practical Instructions on the Formation, Stocking, and Management, in all Seasons, of Collections of River Animals and Plants. Fcap. 8vo, cloth gilt, Illustrated, 2s.

THE BOOK OF THE MARINE AQUARIUM. Practical Instructions on the Formation, Stocking, and Management, in all Seasons, of Collections of Marine Animals and Plants. Fcap. 8vo, cloth gilt, Illustrated, 2s.

THE TOWN GARDEN. A Manual for the Successful Management of City and Suburban Gardens. Second Edition, much enlarged. Fcap. 8vo, cloth, with Illustrations, 3s. 6d.

PROFITABLE GARDENING. A Practical Guide to the Culture of Vegetables, Fruits, and other useful Out-door Garden Products. Intended for the use of Amateurs, Gentlemen's Gardeners, Allottees, and Growers for Market. Small 8vo, cloth, 3s. 6d.

London: GROOMBRIDGE & SONS, 5, Paternoster Row.

NEW BOOKS AND NEW EDITIONS.

THE TEMPLE ANECDOTES,

BY

Ralph Temple

Chandos Temple

NEW WORK BY MORIER EVANS.

Post 8vo, cloth, nearly ready,

SPECULATIVE NOTES

AND

NOTES ON SPECULATION,

IDEAL AND REAL.

BY D. MORIER EVANS,

Author of "Facts, Failures, and Frauds," "History of the Commercial Crisis," etc. etc.

NEW NOVEL BY THOMAS MILLER.

2 vols. post 8vo, cloth, price 21s.

DOROTHY DOVEDALE'S TRIALS,

BY THOMAS MILLER,

Author of "Royston Gower," "Fair Rosamond," "Lady Jane Grey," and "Gideon Giles."

DR. PHIPSON'S NEW WORK.

Small post 8vo, cloth, with Illustrations, nearly ready,

THE

UTILIZATION OF MINUTE LIFE

AND

LOWER ORGANISMS.

Being Practical Studies on Insects, Crustacea, Mollusca, Worms, Polyps, Infusoria, and Sponges.

BY DR. T. L. PHIPSON, F.C.S., London.

London: GROOMBRIDGE & SONS, 5, Paternoster Row.

www.ingramcontent.com/pod-product-compliance
Lightning Source LLC
Chambersburg PA
CBHW022050230426
43672CB00008B/1124